**McGraw-Hill Education**

# Geometry
## Review and Workbook

**McGraw-Hill Education**

# Geometry
# Review and Workbook

Carolyn Catapano Wheater

New York  Chicago  San Francisco  Athens  London  Madrid
Mexico City  Milan  New Delhi  Singapore  Sydney  Toronto

1  2  3  4  5  6  7  8  9  LHS  23  22  21  20  19  18

ISBN      978-1-260-12890-1
MHID      1-260-12890-3

e-ISBN    978-1-260-12891-8
e-MHID    1-260-12891-1

McGraw-Hill products are available at special quantity discounts to use as premiums and sales promotions or for use in corporate training programs. To contact a representative, please visit the Contact Us pages at www.mhprofessional.com.

# Contents

`CHAPTER 12`

# Circles

`CHAPTER 13`

# Conic Sections

`REVIEW 6`

# Circles and Conics

`CHAPTER 14`

# Three-Dimensional Figures

REVIEW 7
# Three-Dimensional Figures

McGraw-Hill Education

# Geometry
## Review and Workbook

# Pretest

*IDK any of this causing me to leave this section blank for later review.*

This pretest is an opportunity to remind yourself of bits of geometry that you already know, to identify areas of the curriculum that may be totally new to you, and generally help you to start thinking about geometry in new ways. Answer all the questions, and try to express your thinking as clearly as you can, even if you're not completely certain of the answer.

1. What is the adjective applied to lines, rays, or segments that meet to form a right angle?

2. Use a compass and a straightedge to construct the bisector of ∠*ABC*.

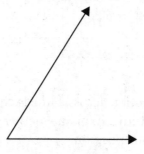

3. Explain the basis of the construction of a line parallel to a given line through a given point.

4. Construct a square inscribed in the given circle.

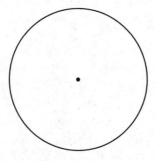

5. If it is true that John plays football, but not true that Elizabeth is a nurse, is it true that John plays football and Elizabeth is a nurse? Explain.

6. Write the converse of *If you study daily, then you will score well on tests.*

7. Don argues that crackers are better than nothing and nothing is better than ice cream, so crackers are better than ice cream. Is his argument valid? Why?

8. ∠A is supplementary to ∠B and ∠B is supplementary to ∠C. Prove that ∠A ≅ ∠C.

9. Draw the image of parallelogram *ABCD* under a translation 2 units right and 3 units down.

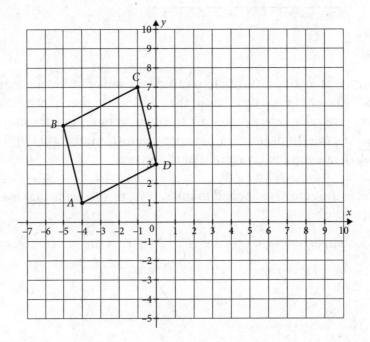

10. Jessica says that a reflection across a line is equivalent to a rotation of 180° around a point on that line. Explain whether or not you agree and give an example to support your position.

**11.** Quadrilateral *ABCD* is a rectangle. Quadrilateral *A′B′C′D′* is the image of *ABCD* under a composition of two transformations. Identify what the two transformations might be, and give specific details.

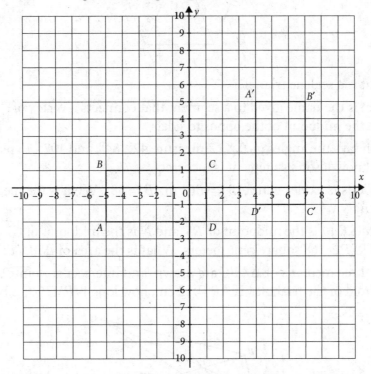

**12.** What information not already indicated on the diagram would you need in order to prove these two triangles congruent by SAS? Write the congruence statement.

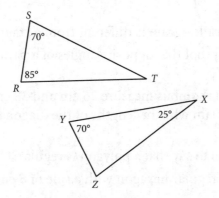

**13.** Triangles can be proven congruent by SAS but not by SSA. Explain why it is possible to prove triangles congruent by ASA or AAS.

**14.** Use transformations to explain how to identify corresponding parts of congruent triangles.

**15.** To prove $\triangle ABF \cong \triangle EDG$, another pair of triangles must be proven congruent first. Which triangles must be proven congruent first and what corresponding parts will help prove $\triangle ABF \cong EDG$?

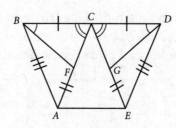

**16.** What is a scalene triangle?

**17.** What is the name given to a segment that connects a vertex of a triangle with the midpoint of the opposite side?

**18.** Jason says the angles in $\triangle XYZ$ measure 42°, 84°, and 107°. Explain how you know Jason's statement is incorrect.

**19.** $\triangle ABC$ is isosceles with $\overline{AB} \cong \overline{BC}$. $m\angle B = 38°$. What is the measurement of $\angle A$?

**20.** In $\triangle RST$. $M$ is the midpoint of $\overline{RS}$ and $N$ is the midpoint of $\overline{ST}$. The area of $\triangle RST$ is 368 square centimeters. What is the area of $\triangle MNT$?

**21.** The three medians of $\triangle ABC$ are drawn, and intersect at point $P$, the centroid of the triangle. If $AP = 12$ cm, how long is $PM$?

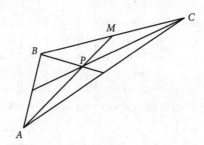

**22.** Explain how a parallelogram is different from a trapezoid.

**23.** Give an informal proof that opposite angles of a parallelogram are congruent.

**24.** The diagonals of a rhombus measure 26 cm and 38 cm. Find the area of the rhombus, and explain why the lengths of the diagonals give you enough information.

**25.** What does it mean to say that a polygon is regular?

**26.** Find the area of a regular hexagon with a side of 6 cm and apothem of $6\sqrt{3}$ cm.

**27.** What is meant by the center of a dilation?

**28.** Draw the image of line segment $\overline{PQ}$ under a dilation by a factor of 1.5 centered at (0, 0).

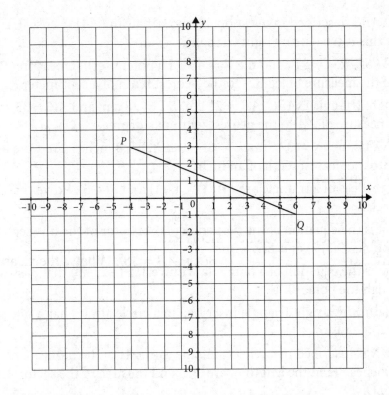

29. Use dilations to explain what it means for two polygons to be similar.

30. How do the lengths of corresponding segments of similar figures compare?

31. $\triangle ABC$ has m$\angle A = 47°$ and m$\angle B = 51°$. $\triangle RST$ has m$\angle R = 51°$ and m$\angle T = 82°$. Are the triangles similar? If so, write a similarity statement. If not, explain why.

32. $\triangle XYZ \sim \triangle MNO$. $YZ = 14$ cm, $NO = 21$ cm, and $XZ = 20$ cm. Find the length of $\overline{MO}$.

33. In $\triangle RST$, point $V$ lies on $\overline{RS}$ and point $W$ lies on $\overline{ST}$. Line segment $\overline{VW}$ is parallel to $\overline{RT}$ and measures 2 inches. $RT = 10$ inches and $ST = 15$ inches. Find the length of $\overline{WT}$.

34. The legs of a right triangle measure 6 cm and 8 cm, and the hypotenuse measures 10 cm. Find the area of the triangle and the length of the altitude drawn from the right angle vertex to the hypotenuse.

35. Find the length of the shorter leg of a right triangle, to the nearest tenth of an inch, if the longer leg measures 6.2 inches and the hypotenuse measures 10.7 inches.

36. Classify $\triangle DEF$ as right, acute, or obtuse, if $DE = 4$ cm, $EF = 6$ cm, and $DF = 7$ cm.

37. Right triangle $\triangle ABC$ has right angle at $B$. $\overline{BD}$, the altitude to the hypotenuse, is drawn, and measures 12 inches. Hypotenuse $AC = 25$ inches. Find the measure of $\overline{AD}$.

38. Explain how you can be certain that all right triangles with an acute angle of 22° are similar.

39. $\triangle ABC$ is a right triangle with $\overline{AB} \perp \overline{BC}$. $AB = 5$ and $BC = 12$. Find $\sin(\angle A)$, $\cos(\angle A)$, and $\tan(\angle A)$.

**40.** In $\triangle ABC$ described in question 39, show that sin $(\angle A) = \cos(\angle C)$. Explain why this is true in any right triangle.

**41.** In $\triangle RST$, $\overline{RS} \perp \overline{ST}$, and m$\angle R = 40°$. Hypotenuse $\overline{RT}$ measures 24 cm. Find the measure of leg $\overline{ST}$ to the nearest tenth of a centimeter.

**42.** In right triangle $\triangle ABC$, $AB = 21$ cm, $BC = 28$ cm, and $AC = 35$ cm. Find the measure of $\angle A$ to the nearest tenth of a degree.

**43.** In $\triangle XYZ$, m$\angle X = 37°$, m$\angle Z = 61°$, and $XZ = 8$ inches. Find the length of $\overline{XY}$ to the nearest tenth of an inch.

**44.** The vertices of a triangle are $P(1, 1)$, $Q(7, 1)$ and $R(4, 5)$. Prove that $\triangle PQR$ is isosceles.

**45.** What is the equation of the perpendicular bisector of the line segment with endpoints $D(3, 7)$ and $E(9, -3)$?

**46.** What is the equation of a parabola with vertex $(-2, -3)$ that passes through the point $(0, 1)$?

**47.** Explain why a right triangle inscribed in a circle always has a diameter as its hypotenuse.

**48.** $\angle LMN$ is inscribed in circle $O$, and measures $41°$. The radius of circle $O$ is 4 inches. Find the length of arc $\overset{\frown}{LN}$ and the area of the sector defined by $\angle LON$.

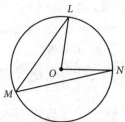

**49.** Elena designed earrings in geometric shapes, including cones and pyramids, and cast them in metal. Because she knows that customers were concerned about earrings being heavy, she wants to be able to estimate the weight of each earring. One of her designs is a cone with a diameter of 1 cm and a height of 2 cm cast in silver, which has a density of 10.49 grams per cubic centimeter. Find the mass of the cone, to the nearest tenth of a gram.

**50.** If a rectangle 3 inches wide and 5 inches long were rotated around its longer edge, what would be the volume of the resulting cylinder?

## CHAPTER 1

# Definitions and Construction Strategies

Every subject of study needs a place to start. The geometry in this course will use many ideas you've learned earlier, but the way in which you examine them, investigate them, challenge them, and apply them may be new. That new approach needs some essential tools: clear understandings of words and their meanings, and the tools to explore new ideas.

## Defining Terms

A geometry course is concerned with investigating, discussing, and proving or disproving ideas about geometric figures. Conversations like that require a shared vocabulary and a clear sense of what the words we use really mean. In previous math classes, you've learned the names of many different lines and shapes and used commonly understood words to describe them.

Geometry accepts the words point, line, and plane as *undefined* terms, but we have to try to describe them to make sure we share the same images. A **point** is often thought of as a tiny dot, but a better image is simply a location. A point takes up no space. A **line** is a collection of points that are side by side, going on forever. A line has infinite length, but no width or height, because it's one point wide and one point high, and points take up no space. All lines are straight. They do not bend or curve. A **plane** is a set of points that align as a flat surface, with infinite length and infinite width, but no height.

Other terms related to lines that you will find useful are:

▶ **Line segment**   A part of a line that includes two points on the line, called *endpoints*, and all the points on the line between them.

▶ **Ray**   A part of a line that includes one point on the line, called the *endpoint*, and all the points on the line to one side of the endpoint.

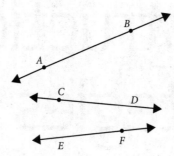

Line $\overleftrightarrow{AB}$, ray $\overrightarrow{CD}$, and ray $\overrightarrow{FE}$. $\overleftrightarrow{AB}$ is identical to $\overleftrightarrow{BA}$, but direction is significant for rays.

▶ **Perpendicular lines**   Two lines that intersect at right angles.

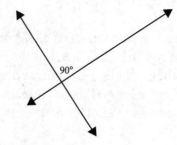

Perpendicular lines intersect to form right angles.

▶ **Parallel lines**   Two lines on the same plane that never intersect.

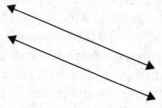

Parallel lines are always the same distance apart and never cross.

▶ **Distance between two points on a line**   The length of the shortest path between them. It is the length of the line segment that has those endpoints. To find that length, you need to establish a coordinate system on the line, or turn it into a number line. If $a$ and $b$ are the coordinates of the endpoints, then the distance between the points is $|a - b| = |b - a|$.

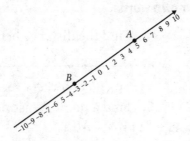

$$AB = BA = |5 - (-3)| = |-3 - 5| = 8$$

▶ **Angle**  The figure formed by two rays with the same endpoint. The common endpoint is the *vertex* of the angle, and the two rays are the *sides* of the angle.

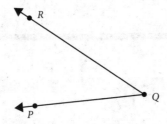

$\angle RQP$ or $\angle PQR$. Point $Q$ is the vertex. The sides are $\overrightarrow{QR}$ and $\overrightarrow{QP}$.

▶ **Circle**  The set of all points in a plane at a common distance, called the *radius*, from a fixed point called the *center*. A portion of the circle between two points on the circle is called an *arc*.

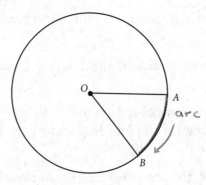

Circle $O$ with radii $\overline{OA}$ and $\overline{OB}$ and arc $\overset{\frown}{AB}$.

▶ **Distance around a circular arc** (or **arc length**)  A fraction of the circumference of the circle. The length of arc $\overset{\frown}{AB}$ in circle $O$ is proportional to the measure of angle $\angle AOB$.

## EXERCISE 1.1

**These exercises focus on the meaning of terms from geometry and the notations we use to communicate them. Answer each question in your own words.**

1.  What is the difference between a line and a segment? *Collection of points going on indefinitely v.s definitely*

2.  How is a ray different from a line and a segment? *A ray has 1 endpoint, line is none, seg. is 1*

3.  Does the picture below show $\overrightarrow{RT}$ or $\overrightarrow{TR}$? Explain. *The picture RT + TR Both also have the same explain the same thing, both have the with two Same space end + distance points being shown.*

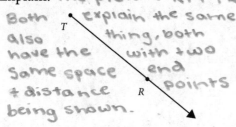

4.  Find the distance between $M$ and $N$.

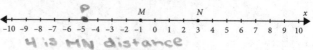

*4 is MN distance*

5.  Find a point $P$, distinct from $M$ or $N$, so that $MN = MP$. *PM will be at point -5*

6.  Explain why the center of a circle is not part of the circle. *The circle is just describe w/the round line!*

7.  Circle $O$ has a circumference of 252 cm. Radii $OA$ and $OB$ are drawn, and $\angle AOB$ is a right angle. What is the length of $\overset{\frown}{AB}$ in centimeters? *AB is 63 cm ∠AOB is proportional AB*

# Constructing a Copy

When you **construct** a geometric figure, you use a compass and straightedge to mark, measure, and connect. The two fundamental constructions are copying a line segment and copying an angle.

▶ **Copy a line segment** In this most basic construction, you use a straightedge to draw a line, and a compass to measure the given segment and mark an equal length on your line:

Given line segment $\overline{AB}$, use a straightedge to draw a new line near $\overleftrightarrow{AB}$.

↓

Choose a point on the new line. Label it $A'$.

↓

Place your compass at point $A$ on the given line. Open the compass until the pencil is on point $B$. (Measure the line segment.)

↓

Holding the compass carefully so that the setting does not change, move the point to $A'$.

↓

Mark an arc that crosses the new line. Label the point of intersection $B'$.

↓

Line $\overline{A'B'}$ is a copy of $\overline{AB}$.

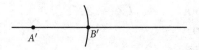

**Copying a segment**

▶ **Copy an angle**    Copying an angle also involves using the compass to measure the opening between the sides and duplicate it for the new angle. Before doing that, however, you'll scribe an arc to ensure that you're measuring the angles in corresponding spots.

Given ∠P, draw a line nearby and choose a point $P'$ to be the image of P.

↓

Place your compass point at the vertex P and scribe an arc that crosses both sides of the angle. Label the intersections Q and R.

↓

Holding the compass carefully so that the setting does not change, move to $P'$ and scribe the same arc. The image angle has only one side now, so there will be only one intersection with the arc. Call it $R'$.

↓

Return to ∠P and place the point of the compass at R where the arc crosses one side of the angle. Open the compass until the pencil touches Q, the intersection of the arc and the other side. (Measure the angle.)

↓

Holding the compass carefully so that the setting does not change, move the point to $R'$. Scribe an arc that crosses the existing arc. Label the intersection $Q'$.

↓

Connect $Q'$ to vertex $P'$ to create the other side of the angle.

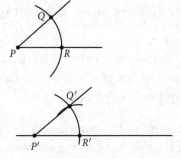

**Copying an angle**

## EXERCISE 1.2

**These exercises ask you to think about the steps in a construction and why they work. Answer each question in your own words. Perform the construction if it helps you form your answer.**

1. When you are copying a segment, which tool (compass or straightedge) should you use to make sure the copy is the same length as the given segment?

2. Copy segment $\overline{PR}$.

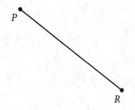

3. In the "copy an angle" construction in the lesson above, why must $\overset{\frown}{QR}$ and $\overset{\frown}{Q'R'}$ have the same radius?

4. Copy $\angle XYZ$.

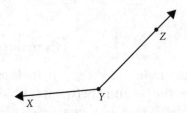

5. If you were given $\angle XYZ$ above, could you create a copy with its vertex at $Z$? Explain.

6. Use the "copy a segment" and "copy an angle" constructions to create an isosceles triangle that contains an angle the size of $\angle ABC$ and has a side the length of $\overline{PQ}$.

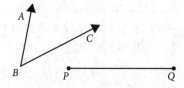

# Constructing a Bisector

The other fundamental constructions are bisecting a segment and bisecting an angle. When combined with the copying constructions, these bisections allow you to perform many different tasks. To **bisect** is to cut into two pieces of equal size.

▶ **Bisect a line segment**   This construction locates two points that lie on the perpendicular bisector of the given segment and connects them to cut the segment into two segments of equal length.

Given $\overline{AB}$, place the point of the compass at $A$, and open the compass until the pencil is on $B$. Scribe an arc above and below $\overline{AB}$.

(The opening of the compass only needs to be _more_ than halfway
from $A$ to $B$, but setting it to the length of $\overline{AB}$ is convenient.)

↓

Place the point of the compass at $B$ and open to put the
pencil on $A$. Scribe an arc above and below.

↓

Label the intersections of the two arcs $P$ and $Q$. Draw $\overleftrightarrow{PQ}$. The point
where $\overleftrightarrow{PQ}$ intersects $\overline{AB}$ is the midpoint of $\overline{AB}$. $\overleftrightarrow{PQ}$ bisects $\overline{AB}$.

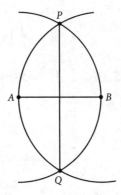

▶ **Bisect an angle**   This construction first scribes an arc to locate a
point on each side equidistant from the vertex. Then it finds a point in
the interior of the angle that is equidistant from those starting points.
Connecting that to the vertex divides the angle into two angles of
equal size.

Given $\angle P$, place the point of the compass at $P$, open to any
convenient setting, and scribe an arc that crosses both
sides of the angle. Label the intersections $A$ and $B$.

↓

Move the compass point to $A$, and open until the pencil
is on $B$. Scribe an arc in the interior of the angle.

↓

Move the compass point to $B$, and open until the pencil is on $A$.
Scribe an arc in the interior of the angle. Label
the intersection of the two arcs $Q$.

↓

Ray $\overrightarrow{PQ}$ bisects $\angle APB$.

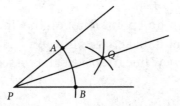

## EXERCISE 1.3

**The exercises below ask you to perform some constructions and to think about why the steps in the constructions have the desired effect. Answer the questions in your own words. Perform the construction if it helps you formulate your response.**

1. When you bisect a segment as shown above, how do you know that the distance *PA* is equal to the distance *PB*?

2. Bisect $\overline{PQ}$.

3. Bisect two sides of $\triangle ABC$. Label the midpoints *M* and *N*. Draw $\overline{MN}$.

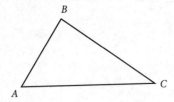

4. Compare $\overline{MN}$ to the side you did not bisect. What do you notice?

5. Bisect $\angle XYZ$.

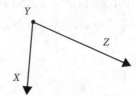

6. Bisect all three angles of $\triangle RST$.

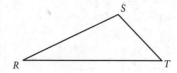

7. What do you notice about the angle bisectors you created in $\triangle RST$?

# Constructing Parallels and Perpendiculars

Constructing a line parallel to a given line, through a given point, is an application of copying an angle. You'll draw a line through the given point and the given line, creating a transversal. Then you'll copy the angle formed by the given line and the transversal. Because these corresponding angles are congruent, the lines are parallel.

▶ **Construct a line parallel to a given line**  This construction uses the given point to create a transversal, which forms an angle with the given line. That angle is copied with the vertex at the given point. Because the two corresponding angles are congruent, the lines cut by the transversal are parallel.

Given line $\overleftrightarrow{AB}$ and point *P* not on the line, draw a line through point *P* that crosses line $\overleftrightarrow{AB}$. Label the point where this new line (transversal) crosses $\overleftrightarrow{AB}$ as *Q*.

↓

With the compass point at *Q*, scribe an arc of any convenient size that crosses both $\overleftrightarrow{AB}$ and $\overleftrightarrow{PQ}$. Label the points of intersection *R* and *S*.

↓

Hold the same setting on the compass, move the point to $P$ and scribe the same arc. Label the point where the arc crosses $\overleftrightarrow{PQ}$ as point $T$.

↓

Use the compass to measure from $R$ to $S$, move the compass with that setting to $T$, and scribe an arc. Label the intersection of the two arcs $V$.

↓

Draw line $\overleftrightarrow{PV}$. $\overleftrightarrow{PV} \parallel \overleftrightarrow{AB}$.

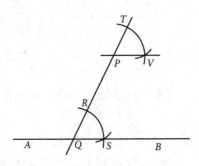

▶ **Construct a line perpendicular to a given line**    To construct a line perpendicular to a given line through a particular point on the line, you'll borrow the "bisect a line segment" construction. That construction creates the perpendicular bisector of the segment so it gives you the perpendicular you want. You just need to take one extra step first.

With the point of the compass on point $P$ on the given line $\overleftrightarrow{AB}$ and any convenient radius, scribe a circle. Label the points where the circle intersects $\overleftrightarrow{AB}$ as point $C$ and point $D$. Point $P$ is the midpoint of segment $\overline{CD}$.

↓

Construct the bisector of $\overline{CD}$. With the compass set to the length of $\overline{CD}$, scribe an arc from $C$ that goes both above and below $\overline{CD}$. Repeat from $D$. Label the intersections of the two arcs as point $E$ and point $F$.

↓

Draw line $\overleftrightarrow{EF}$. (Notice it passes through $P$.) $\overleftrightarrow{EF} \perp \overline{CD}$ and $\overleftrightarrow{EF} \perp \overleftrightarrow{AB}$.

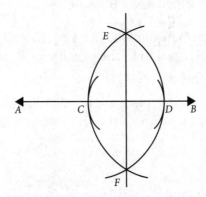

## EXERCISE 1.4

**Complete the constructions requested below, and for questions that ask you to analyze a construction, explain in your own words as clearly as you can.**

1. In the example of constructing parallel lines above, ∠RQS is copied to ∠TPV. Would the construction still have produced parallel lines if ∠RQA had been copied to an angle with vertex P? Explain.

2. Construct a line parallel to $\overline{AB}$ through point Z.

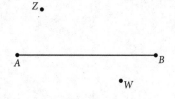

3. Construct a line parallel to $\overline{AB}$ through point W. Is the new line parallel to the line constructed in question 2? Explain.

4. When constructing a perpendicular to a line through a point on the line, you can use "any convenient radius" to define the segment to bisect. Why is that possible? Why don't we need to specify a certain radius?

5. Construct a line perpendicular to $\overline{MN}$ through point P.

6. Locate a point Q on the perpendicular so that PQ = PN.

7. Bisect $\overline{MP}$. Label the midpoint R. Draw $\overline{RQ}$. What kind of triangle is ΔRPQ?

# Constructing Regular Polygons

In a later chapter, you'll learn constructions that inscribe regular polygons in a circle, but for now, there are two constructions that create regular polygons using the construction of a perpendicular and the compass as a measuring tool.

▶ **Construct an equilateral triangle**   This construction scribes two arcs, each with a radius equal to the given segment. The point at which the arcs intersect is point C for which AB = BC = AC.

Given a segment $\overline{AB}$ to be one side of an equilateral triangle, place the compass point at A, open until the pencil is at B, and scribe an arc.

↓

With the same setting, place the point at B and scribe an arc.
Label the point at which the arcs intersect as point C.

↓

Draw $\overline{AC}$ and $\overline{BC}$. $\triangle ABC$ is an equilateral triangle.

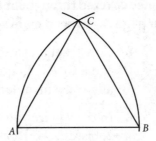

▶ **Construct a square**   A square is a rectangle, so it must have a right angle at each vertex, and its four sides must all be of equal length. You'll construct perpendiculars to ensure the right angles, and then use the compass as a measuring tool to guarantee sides of equal length.

Given $\overline{AB}$ to be one side of a square, extend through $A$, and construct a line perpendicular to $\overline{AB}$ through $A$.

↓

Scribe a circle of any convenient radius centered at $A$ to define a segment that has $A$ as its midpoint.

↓

Construct the perpendicular bisector of that segment to create a line perpendicular to $\overline{AB}$ through $A$.

↓

Extend $\overline{AB}$ through $B$ and repeat the steps to define a segment that has $B$ as its midpoint, and construct the perpendicular bisector.

↓

When perpendiculars have been created at both $A$ and $B$, set the compass to the length of $\overline{AB}$. With point at $A$, scribe an arc that intersects the line perpendicular to $\overline{AB}$ at $A$. Label the intersection point $D$.

↓

With the same compass setting and compass point at $B$, scribe an arc that intersects the line perpendicular to $\overline{AB}$ at $B$. Label the intersection $C$.

↓

Draw $\overline{DC}$. $ABCD$ is a square.

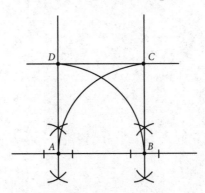

## EXERCISE 1.5

**These exercises bring together many of the techniques covered throughout the chapter. Complete each construction. Copy the given information to another page if you need more space.**

**1.** Construct an equilateral triangle with $\overline{RT}$ as one side.

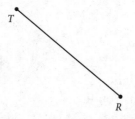

**2.** Construct a square with side $\overline{WX}$.

**3.** Construct an isosceles triangle with $\overline{WX}$ as its base. (The base is the side not equal to the other two. Choose any convenient length for the two equal sides.)

**4.** Construct an isosceles triangle with two sides equal to $\overline{RT}$ in question 1 above. (Choose any convenient length for the base.)

**5.** In the square you constructed in question 2, bisect $\angle W$. Does the angle bisector pass through another vertex of the square? Does it bisect that angle as well? Explain.

**6.** In the equilateral triangle you constructed in question 1, bisect $\angle R$. Does that angle bisector also bisect the opposite side?

## CHAPTER 2

# Logic and Proof

One of the principal differences between a course in geometry and the geometric concepts covered in earlier courses is an emphasis on justifying conclusions and proving them to be true. To make those arguments successfully, you should have a clear understanding of the logic behind them. Geometry builds these arguments with **statements**, which are sentences that can be deemed true or false.

## And, Or, and If-Then

Sometimes it's easy to decide if a simple statement is true or false. "Sacramento is the capital of California" is a true statement. "Miami is the capital of Florida" is not true. You can look it up. Convincing others that a geometric statement is true, however, may take an argument made up several true statements that connect to one another in a logical fashion. Common ways to connect statements are with the word *and* or the word *or*. The *and* statements are called **conjunctions** and the *or* statements are called **disjunctions**. It's also possible to talk about the negation or opposite of a statement:

▶ The **negation** or opposite of a statement is false if the original statement is true and true if the original statement is false. If the original statement is symbolized by $p$, the negation is $\sim p$.

The negation of the statement "Denver is in Colorado" is "Denver is not in Colorado." The first statement is true, and the second false.

The negation of the false statement "Mars has twelve moons" is the true statement "Mars does not have twelve moons."

▶ When two statements are joined with the word *and*, both must be true or the conjunction is false.

The statement "Sacramento is the capital of California and Miami is the capital of Florida" can't be labeled true, because only part of it is true.

▶ Using *and* demands that both parts of the conjunction be true. The symbolic representation of a conjunction is $p \wedge q$.

"Paris is north of the equator and Capetown is south of the equator" is a true conjunction.

▶ An *or* statement, or disjunction, is true if either or both statements are true. Saying Sacramento is the capital of California or Miami is the capital of Florida doesn't sound false because the word *or* says either one. One or the other works. What about both? In general, *or* is assumed to be inclusive, so it says one or the other or both. The symbolic representation of a disjunction is $p \vee q$.

"Grains provide carbohydrates or fruits contain natural sugars" is true because both are true. "Tokyo is the most populous city in the world or Mumbai is in China" is true as of this writing, because, although Mumbai is not in China, Tokyo is currently the most populous city. If the population of Tokyo were to decline, so that it lost that first-place ranking, then both pieces would be false and the disjunction would be false.

When we mean one or the other but not both, we're using an **exclusive or**, usually abbreviated as xor.

The other common way to connect two statements, and the one most commonly used in arguments, is the **if-then**, or **conditional**, statement. It can help to think of a conditional statement like a promise or a contract. For example: If you clean the garage, then I will wash the car. The *if* section, or **hypothesis**, is true, and the conclusion, which follows *then*, is true, and everything is fine. You clean the garage. I wash the car. We have met the terms of our deal. But if you clean the garage and then find I haven't washed the car, you are entitled to feel I've broken the deal. What if, after agreeing to the deal, you don't clean the garage? If I don't wash the car, or even if I do, you can't complain. You didn't do your part. An **if-then** or **conditional** statement is only false when a true hypothesis leads to a false conclusion. The symbolic representation of a conditional is $p \rightarrow q$.

Consider these four statements:

| | | |
|---|---|---|
| If $2 < 5$, then $2 + 4 < 5 + 4$ | If True, then True | TRUE |
| If $2 < 5$, then $2 + 4 < 5$ | If True, then False | FALSE |
| If $5 < 3$, then $5 - 4 < 3$ | If False, then True | TRUE |
| If $5 < 3$, then $5 + 7 < 3 + 7$ | If False, then False | TRUE |

Conditional statements don't always appear in *if-then* form. Rewriting them requires care to be sure the hypothesis and conclusion aren't confused. *A triangle is isosceles if it has two sides of equal length* translates to *if a triangle has two sides of equal length, then the triangle is isosceles.*

## EXERCISE 2.1

**Use the principles of logic in this section to construct responses to these exercises.**

1. Let *p* be the statement "$\triangle ABC$ is a right triangle." Let *q* be "$\triangle ABC$ is an isosceles triangle." Let *r* be "$\triangle ABC$ is an equilateral triangle." Under what circumstances is $p \wedge q$ true? Under what circumstances is $p \wedge r$ true?

2. Suppose *ABCD* is a quadrilateral with four congruent sides. "*ABCD* is a rhombus or a square" is a simplified version of the long form statement "*ABCD* is a rhombus or *ABCD* is a square." Is the statement "*ABCD* is a rhombus or a square" a true statement? Explain.

3. The negation of "*ABCD* is a rhombus" is "*ABCD* is not a rhombus," but what is the negation of "*ABCD* is a rhombus or a square"? Explain your thinking, and write the negation in the long form.

4. Assume $\triangle ABC$ is equilateral. Label this conditional statement true or false: "If $\triangle ABC$ is not equilateral, then it is a right triangle." Explain your thinking.

5. Assume $\triangle ABC$ is equilateral. Label this conditional statement true or false: "If $\triangle ABC$ is equilateral, then it is not isosceles." Explain your thinking.

6. Rewrite in *if-then* form: "Lines are perpendicular when they intersect to form a right angle."

7. Give the two conditionals that form the statement: "A triangle is equiangular if and only if it is equilateral."

# Truth Tables

A **truth table** is a tool for organizing all the possible truth values for a statement or for deciding if a pattern of argument is valid. The table is built by using *p*, *q*, and other lowercase letters to represent the individual statements, ~*p*, ~*q*, and so on to represent negations, and the symbols $\wedge$ for *and*, $\vee$ for *or*, and $\rightarrow$ for *if-then*. There is a line for each possible combination of true and false statements; for example, if there are two statements, *p* and *q*, there would be four lines: *TT*, *TF*, *FT*, and *FF*.

### Conjunction

| *p* | *q* | *p* $\wedge$ *q* |
|:---:|:---:|:---:|
| T | T | T |
| T | F | F |
| F | T | F |
| F | F | F |

**Disjunction**

| p | q | p ∨ q |
|---|---|---|
| T | T | T |
| T | F | T |
| F | T | T |
| F | F | F |

**Conditional**

| p | q | p → q |
|---|---|---|
| T | T | T |
| T | F | F |
| F | T | T |
| F | F | T |

Truth tables could be used to summarize possible truth values for each of the types of statements in the last lesson.

The more interesting use of truth tables is to help to determine if a particular line of reasoning is a valid argument. A pattern of reasoning is considered valid if it is a **tautology**, that is, if the result is true for every line of the table.

EXAMPLE

▶ Suppose we argue "If a triangle has three sides of equal length, then it is equilateral, and △ABC has three sides of equal length. Therefore, △ABC is equilateral."

▶ Let p represent "a triangle has three sides of equal length" and q represent "a triangle is equilateral." Then the argument is [(p→q) ∧ p]→q. The table would look like this:

| p | q | p → q | (p → q) ∧ p | [(p→q) ∧ p]→q |
|---|---|---|---|---|
| T | T | T | T ∧ T<br>T | T → T<br>T |
| T | F | F | F ∧ T<br>F | F → F<br>T |
| F | T | T | T ∧ F<br>F | F → T<br>T |
| F | F | T | T ∧ F<br>F | F → F<br>T |

## EXERCISE 2.2

**Use the information in this section to construct truth tables. Use the truth tables to help you make judgments about the truth of the statements.**

1. Construct a truth table for $q \rightarrow p$. Compare your result to the table above for $p \rightarrow q$. What differences do you see?

2. In the example above we saw that $[(p \rightarrow q) \wedge p] \rightarrow q$ is a tautology. Is $[(p \rightarrow q) \wedge q] \rightarrow p$ also a tautology?

3. Construct a truth table for $\sim p \rightarrow \sim q$ and one for $\sim q \rightarrow \sim p$. Compare them to the tables for $p \rightarrow q$ and $q \rightarrow p$ and discuss the similarities and differences you observe.

4. Identify the simple sentences from which the compound sentence below is formed. Define a symbol for each simple sentence and write the compound sentence in symbolic form.

   If $ABCD$ is a parallelogram and its diagonals are congruent, then $ABCD$ is a rectangle or a square.

5. Suppose $p$: $\triangle RST$ is a right triangle, $q$: $RS = ST$, and $r$: m$\angle R =$ m$\angle T$. Write $(p \wedge q) \rightarrow r$ in words.

6. If you wanted to build a truth table for the statement in question 5, which is built from three simple statements, how many lines would the table need to cover all cases? Complete the list below. How many lines would be needed if the statement involved four simple statements?

   | p | q | r |
   |---|---|---|
   | T | T | T |

7. If you were to create a truth table for an *exclusive or*, how would it be different from the truth table for *or*?

# Converses, Inverses, and Contrapositives

Every conditional statement has three related statements, formed either from the same statements, $p$ and $q$, or their negations, $\sim p$ and $\sim q$. If $p \rightarrow q$ is a conditional statement, the converse of $p \rightarrow q$ is $q \rightarrow p$, the inverse of $p \rightarrow q$ is $\sim p \rightarrow \sim q$, and the contrapositive of $p \rightarrow q$ is $\sim q \rightarrow \sim p$.

| | |
|---|---|
| **Conditional** | If a triangle is equilateral, then it is isosceles. |
| **Converse** | If a triangle is isosceles, then it is equilateral. |
| **Inverse** | If a triangle is not equilateral, then it is not isosceles. |
| **Contrapositive** | If a triangle is not isosceles, then it is not equilateral. |

As you saw in the exercises in the last lesson, and perhaps here in these example sentences, the converse and inverse may or may not have the same truth value as the original conditional, but the contrapositive always matches the original. In this example, it's true that an equilateral triangle is isosceles, but having two sides of equal length doesn't guarantee the third is the same length. The converse is not equivalent to the conditional. But if a triangle does not have two sides of equal length, it can't have three sides of equal length. The contrapositive is equivalent to the conditional.

When the original conditional and the converse are joined with an *and*, the compound statement is called a **biconditional**. Symbolically, a biconditional is written as $(p \rightarrow q) \wedge (q \rightarrow p)$ or the shorter form $p \leftrightarrow q$. When written in words, it is generally condensed to *if and only if*, as in "a triangle is equilateral *if and only if* it is isosceles."

A biconditional is true when both statements have the same truth value, both true or both false.

> *If and only if* is sometimes abbreviated **iff.**

| $p$ | $q$ | $p \rightarrow q$ | $q \rightarrow p$ | $(p \rightarrow q) \wedge (q \rightarrow p)\ p \leftrightarrow q$ |
|---|---|---|---|---|
| T | T | T | T | $T \wedge T$ <br> T |
| T | F | F | T | $F \wedge T$ <br> F |
| F | T | T | F | $T \wedge F$ <br> F |
| F | F | T | T | $T \wedge T$ <br> T |

## EXERCISE 2.3

**The exercises below deal with the converse, inverse, and contrapositive of conditional statements, and how their truth values are related. Answer each question by writing the requested statement in words or symbols, as directed.**

1. Write the converse of "If a quadrilateral has two pairs of parallel sides, then it is a parallelogram."

2. Write the inverse of "If a quadrilateral is a parallelogram, then its diagonals bisect one another."

3. Write the contrapositive of "If the diagonals of a parallelogram are congruent, then the parallelogram is a rectangle."

4. Suppose $p \leftrightarrow q$ is the statement "Two lines are perpendicular if and only if they intersect to form right angles. Write out statement $p$ and statement $q$.

5. What is the converse in symbolic form of a statement of the form $\sim p \rightarrow q$? What is the inverse?

6. What is the converse of the inverse of $p \rightarrow q$?

7. What is the inverse of the converse of $\sim q \rightarrow \sim p$?

# Arguments

When you want to convince someone of the truth of a statement, it's important that your argument is based on sound logic as well as true statements. It's true that equilateral triangles are isosceles, and it's true that the hypotenuse is the longest side of a right triangle, but if you don't have a sensible way to connect those ideas, you can't prove anything.

Two patterns of reasoning are common. One is called **detachment**. You start with a conditional statement; for example, if a triangle is isosceles, then it has two angles of equal size. Follow with a simple statement: $\triangle ABC$ is isosceles. You can then conclude $\triangle ABC$ has two angles of equal size.

A truth table shows that the pattern of reasoning called detachment is a tautology and is therefore a valid argument.

| $p$ | $q$ | $p \to q$ | $(p \to q) \wedge p$ | $[(p \to q) \wedge p] \to q$ |
|---|---|---|---|---|
| T | T | T | T | T → T<br>T |
| T | F | F | F | F → F<br>T |
| F | T | T | F | F → T<br>T |
| F | F | T | F | F → F<br>T |

As long as the second statement matches the hypothesis ($p$) of the conditional, you can decide that $q$, the conclusion, is true. But it doesn't get you very far.

The second common pattern of reasoning allows you to link as many conditionals as necessary, but technically this pattern, called **syllogism**, says if $p \to q$ and $q \to r$, then $p \to r$. If a sequence of conditional statements links conclusion to hypothesis, you can condense the sequence to if the first hypothesis, then the final conclusion:

▶ If quadrilateral $ABCD$ has two pairs of parallel sides, then $ABCD$ is a parallelogram.

If $ABCD$ is a parallelogram, then its diagonals bisect one another.

↓

Therefore, if quadrilateral $ABCD$ has two pairs of parallel sides, then its diagonals bisect one another.

## EXERCISE 2.4

Constructing a valid argument is critical to the proofs on which geometry is built. In each exercise below, carefully determine whether the arguments are valid.

1. Is the argument below valid? Explain your reasoning.

   If a quadrilateral is a rhombus, then it is a parallelogram. *ABCD* is a parallelogram; therefore, *ABCD* is a rhombus.

2. What conclusion can you reasonably draw from the statements below?

   If a triangle is isosceles, then it has two congruent sides. If a triangle has two congruent sides, the angles opposite those sides are congruent. $\triangle RST$ is isosceles.

3. In the argument below, is the conclusion drawn true?

   If a quadrilateral has two pairs of opposite sides congruent, then it is a parallelogram. If a quadrilateral has two pairs of opposite angles congruent, then it is a parallelogram. Therefore, if a quadrilateral has two pairs of opposite sides congruent, then it has two pairs of opposite angles congruent.

4. Use a truth table to show whether the pattern of reasoning below represent a valid argument.

   $[(p \rightarrow q) \wedge (r \rightarrow q)] \rightarrow (p \rightarrow r)$

5. In the argument below, is the conclusion drawn true?

   If a quadrilateral is a parallelogram, then its diagonals bisect one another. If a quadrilateral is a rhombus, then its diagonals bisect one another. Therefore, if a quadrilateral is a parallelogram, then it is a rhombus.

6. Is it possible to reach a conclusion that is true when the structure of your argument is not valid? Explain.

7. If you don't get out of bed, then you will be late for school. If you are late for school, then you will get detention. If you get detention, you will miss the game. You didn't miss the game. What conclusion can you draw?

# Vertical Angles

The undefined terms and the few essential definitions geometry starts with don't generate a lot of ideas that need to be proved. But when two lines cross, four angles are formed, and the relationships among those angles lead to new conjectures. The two lines can be viewed as four rays all starting from the same endpoint. Two rays with the same endpoint that project in opposite directions to form a line are called **opposite rays**. $\overrightarrow{PX}$ and $\overrightarrow{PY}$ are opposite rays, and $\overrightarrow{PA}$ and $\overrightarrow{PB}$ are another pair of opposite rays.

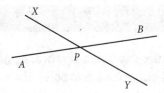

The four angles $\angle APX$, $\angle XPB$, $\angle BPY$, and $\angle YPA$, all share a vertex, $P$, and each pair of adjacent angles shares a side. $\angle APX$ shares side $\overrightarrow{PA}$ with $\angle YPA$ and side $\overrightarrow{PX}$ with $\angle XPB$. Any two adjacent angles in this configuration have one shared side and the remaining sides, called exterior sides, form opposite rays, or a line. Adjacent angles whose exterior sides form a line are called a **linear pair**. Because the opposite rays can be thought of as an angle of 180°, the angles of a linear pair have measures that add to 180°. They are supplementary angles.

▶ Two angles whose measures add to 180° are supplementary.

▶ Two angles whose measures add to 90° are complementary.

▶ Linear pairs are supplementary.

The four angles at the intersection form two pairs of vertical angles. **Vertical angles** are defined as two angles that have a common vertex and whose sides form opposite rays. It may be simpler to say that vertical angles are the pair of angles across the "X" when lines intersect. $\angle APX$ and $\angle BPY$ are vertical angles, and $\angle XPB$ and $\angle YPA$ are vertical angles.

If $\angle XPB$ and $\angle YPA$ are vertical angles, then $\overrightarrow{PX}$ and $\overrightarrow{PY}$ are opposite rays, and $\overrightarrow{PA}$ and $\overrightarrow{PB}$ are opposite rays. If those are opposite rays, then $\angle APX$ and $\angle XPB$ are a linear pair, and $\angle APX$ and $\angle YPA$ are a linear pair. If two angles form a linear pair, the angles are supplementary, which means their measures add to 180°. If $m\angle APX + m\angle XPB = 180°$ and $m\angle APX + m\angle YPA = 180°$, then $m\angle APX + m\angle XPB = m\angle APX + m\angle YPA$. If $m\angle APX$ is subtracted from both sides, $m\angle XPB = m\angle YPA$. If we were to draw a different pair of lines intersecting, the only part of this argument that would change would be the names, so we can confidently make this statement for any pair of vertical angles.

▶ Vertical angles have equal measurements.

▶ If two angles are supplementary to the same angle, they are congruent to each other.

▶ If two angles are complementary to the same angle, they are congruent to each other.

## EXERCISE 2.5

These exercises connect essential geometric structures to basic algebra. Apply the definitions and properties of different types of angles to construct equations and solve them to find angle measures.

1. Name two pairs of vertical angles in the figure below.

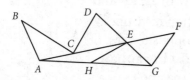

2. Use vertical angles and linear pairs to find the measures of all the angles in the figure below.

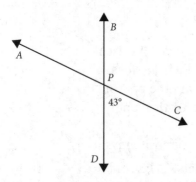

3. $\overrightarrow{PQ}$ bisects $\angle APB$. Find m$\angle QPC$.

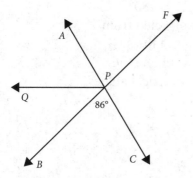

4. Line $\overleftrightarrow{PQ}$ and line $\overleftrightarrow{ST}$ intersect at R. m$\angle PRS$ is twice m$\angle SRQ$. Explain why it is not possible that $\overleftrightarrow{PQ} \perp \overleftrightarrow{ST}$.

5. Line $\overleftrightarrow{AB}$ and line $\overleftrightarrow{CD}$ intersect at E. m$\angle AEC$ $=2(3x + 1)° =$ and m$\angle CEB = (9x − 2)°$. Find the value of $x$.

6. Line $\overleftrightarrow{XY}$ and line $\overleftrightarrow{VW}$ intersect at Z. m$\angle XZV = x°$ and m$\angle XZW = (x − 54)°$. Find m$\angle XZW$.

7. In the figure below, find $x$.

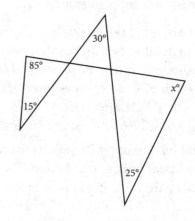

# Parallel Lines

As discussed, parallel lines are defined as lines in the same plane that never intersect. The mention of the lines being in the same plane, or coplanar, distinguishes parallel lines from skew lines. **Skew lines** never intersect but are not coplanar. Railroad tracks are often given as an example of parallel lines. They lie on the same plane and must always be the same distance apart, so they never meet. But signposts or trees along the train's right of way also do not meet the tracks (we hope) yet are not parallel to the tracks. The signpost and the track are skew lines.

A line that crosses two or more lines at different points is called a **transversal**. When a transversal cuts through other lines, a cluster of four angles is created at each intersection. These angles are named according to their location. The angles created when transversal $\overleftrightarrow{EF}$ crosses $\overleftrightarrow{AB}$ are numbered 1 through 4, and the angles created when transversal $\overleftrightarrow{EF}$ crosses $\overleftrightarrow{CD}$ are labeled 5 through 8.

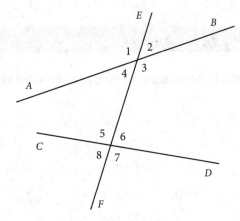

**Corresponding angles** are a pair of angles, one from each intersection, which are in the same relative position in their cluster. ∠1 and ∠5, each in the upper left position, are corresponding angles. ∠2 and ∠6 are a pair of corresponding angles, because each sits in the upper right of its cluster. Other pairs of corresponding angles are ∠4 and ∠8 (lower left) and ∠3 and ∠7 (lower right).

▶ When the two lines cut by a transversal are parallel lines, a pair of corresponding angles has the same measure.

▶ If a pair of lines is cut by a transversal in such a way that corresponding angles are congruent, then the lines are parallel.

**Alternate interior angles** is the name given to two angles, one from each cluster, one from each side of the transversal, both located between the lines. ∠4 and ∠6 are a pair of alternate interior angles, as are ∠3 and ∠5.

▶ If parallel lines are cut by a transversal, alternate interior angles are congruent.

▶ If a pair of lines is cut by a transversal in such a way that alternate interior angles are congruent, then the lines are parallel.

**Alternate exterior angles** are a pair of angles, one from each cluster, one from each side of the transversal, located outside the lines. ∠1 and ∠7 or ∠2 and ∠8 are a pair of alternate exterior angles.

▶ If parallel lines are cut by a transversal, alternate exterior angles are congruent.

▶ If a pair of lines is cut by a transversal in such a way that alternate exterior angles are congruent, then the lines are parallel.

**Consecutive interior angles** describe ∠3 and ∠6 or ∠4 and ∠5. They are between the lines, but on the same side of the transversal.

▶ If parallel lines are cut by a transversal, consecutive interior angles are supplementary.

▶ If a pair of lines is cut by a transversal in such a way that consecutive interior angles are supplementary, then the lines are parallel.

## EXERCISE 2.6

**Use properties of parallel lines, and algebra where necessary, to find the required angle measures.**

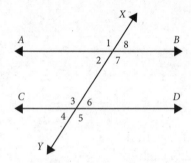

1. In the figure above, $\overleftrightarrow{AB}$ is parallel to $\overleftrightarrow{CD}$. Transversal $\overleftrightarrow{XY}$ is drawn. If m∠1 = 108°, find the measures of all the remaining angles in the figure.

2. ∠1 and ∠3 are corresponding angles. ∠2 and ∠6 are alternate interior angles. Use the postulate above to prove that m∠2 = m∠6.

3. Show that with the previous exercise it is possible to prove m∠4 = m∠8.

4. What conclusion can you draw about the measures of ∠2 and ∠3? Make an argument to support your conclusion.

5. In the figure above, $\overleftrightarrow{AB}$ is parallel to $\overleftrightarrow{CD}$. If $\overleftrightarrow{EF}$ is constructed parallel to $\overleftrightarrow{CD}$, how would you prove that $\overleftrightarrow{EF} \parallel \overleftrightarrow{AB}$?

6. If m∠2 = 2x + 4 and m∠3 = 5x + 1, find the measure of ∠5.

7. If $\overleftrightarrow{WV}$ is added to the figure so that $\overleftrightarrow{WV} \parallel \overleftrightarrow{XY}$, as shown below, which numbered angles are congruent to ∠9?

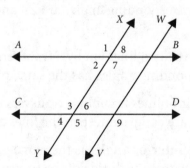

8. If m∠9 = 150°, what is the measure of ∠8?

# Perpendiculars

The definition of perpendiculars talks about lines intersecting to create a right angle. A right angle could also be created by two line segments, or two rays, or by a combination of lines, segments, or rays. In addition, if the intersection creates one right angle, can we be certain that all the angles at that intersection are right angles? It may seem obvious, but you can prove it by using linear pairs and vertical angles.

▶ If two lines intersect to create one right angle, they create four right angles.

▶ All right angles are congruent.

Other statements about perpendicular lines will appear in the exercises for you to prove. These include:

▶ If two lines in the same plane are both perpendicular to the same line, they are parallel to each other.

▶ If a line in the same plane as two parallel lines is perpendicular to one of the parallels, it is perpendicular to the other.

## EXERCISE 2.7

**Use properties of parallel and perpendicular lines to write equations and solve them to find the requested angle measures.**

1. $\overleftrightarrow{AB} \perp \overleftrightarrow{CD}$ at $P$. $\angle APD$ is a right angle. Explain how you know $\angle APC$, $\angle BPC$, and $\angle BPD$ are also right angles.

2. $\overleftrightarrow{AB} \perp \overleftrightarrow{CD}$ at $P$. Ray $\overrightarrow{PQ}$ bisects $\angle APD$ and $\overrightarrow{PR}$ bisects $\angle APC$. What is the measure of $\angle QPR$?

3. $\overleftrightarrow{AB}$, $\overleftrightarrow{CD}$, and $\overleftrightarrow{MN}$ are coplanar. $\overleftrightarrow{AB} \perp \overleftrightarrow{MN}$ and $\overleftrightarrow{CD} \perp \overleftrightarrow{MN}$. Explain how you can conclude that $\overleftrightarrow{AB} \parallel \overleftrightarrow{CD}$.

4. $\overleftrightarrow{PQ}$ and $\overleftrightarrow{RS}$ intersect at $T$. $m\angle PTR = 2x+18$ and $m\angle RTQ = 3x - 13$. Is $\overleftrightarrow{PQ} \perp \overleftrightarrow{RS}$? $\overleftrightarrow{PQ}$ and $\overleftrightarrow{RS}$.

5. $\overleftrightarrow{XY}$, $\overleftrightarrow{RS}$, and $\overleftrightarrow{PQ}$ are coplanar and $\overleftrightarrow{PQ} \parallel \overleftrightarrow{RS}$. If $\overleftrightarrow{XY} \perp \overleftrightarrow{PQ}$, show that $\overleftrightarrow{XY} \perp \overleftrightarrow{RS}$.

6. $\overleftrightarrow{CD} \parallel \overleftrightarrow{EF}$. $\overleftrightarrow{AB}$ intersects $\overleftrightarrow{CD}$ at $P$ and $\overleftrightarrow{EF}$ at $Q$. $m\angle APC = 4x + 18$ and $m\angle EQB = 6x - 18$. Is $\overleftrightarrow{AB} \perp \overleftrightarrow{CD}$? Explain your reasoning.

7. $\overleftrightarrow{RS}$ intersects $\overleftrightarrow{XY}$ at $A$, and $\overleftrightarrow{PQ}$ intersects $\overleftrightarrow{XY}$ at $B$. $m\angle RAX = 5x + 4$ and $m\angle PBY = 6x - 11$. Is $\overleftrightarrow{RS} \perp \overleftrightarrow{XY}$? Is $\overleftrightarrow{PQ} \perp \overleftrightarrow{XY}$? Is $\overleftrightarrow{RS} \parallel \overleftrightarrow{PQ}$? Explain.

# Definitions, Constructions, Logic, and Proof

These review questions bring together the ideas from Chapters 1 and 2 and give you a chance to make sure you understand the vocabulary, have mastered construction techniques, and are learning how to build a logical proof. Answer all the questions, and try to express your thinking as clearly as you can.

1. What does it mean to bisect a segment?

2. Is it possible to bisect a line? Explain.

3. Under what circumstances is a conditional statement of the form $p \rightarrow q$ false?

4. Use a truth table to decide if $\sim p \wedge \sim q$ is the negation of $p \vee q$.

5. Bisect $\angle APB$ using a compass and a straightedge.

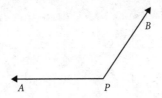

**6.** Let *p* be the statement △*ABC* is a right triangle, and let *q* be the statement *AC* is the longest side of △*ABC*. Write in words: $p \rightarrow (q \vee \sim q)$. Under what circumstance is the statement true? Explain your reasoning.

**7.** In the figure below, $\overline{AB} \parallel \overline{CD}$ and $\overline{XY} \parallel \overline{WV}$. Find the measurement of ∠2 if m∠9 = 117°.

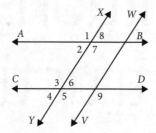

**8.** Use a compass and a straightedge to copy ∠*COB*.

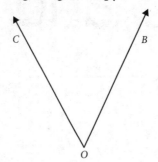

**9.** Is the inverse of a conditional statement $p \rightarrow q$ equivalent to $\sim(p \rightarrow q)$? Explain your reasoning.

**10.** Write the inverse of *If ABCD is a rectangle, then ABCD is a parallelogram.*

**11.** What does it mean to say that lines are perpendicular?

**12.** $\overrightarrow{AB}$ and $\overleftrightarrow{CD}$ are cut by transversal $\overleftrightarrow{XY}$ at points *P* and *Q*, respectively. Is ∠*APX* ≅ ∠*CQX*? What information is important for your decision?

**13.** Write the contrapositive of

*If $\overline{MN}$ connects the midpoint of side $\overline{AB}$ to the midpoint of side $\overline{BC}$ in △ABC, then $\overline{MN}$ is parallel to $\overline{AC}$ and $\overline{MN}$ half as long as $\overline{AC}$.*

**14.** Use a compass and a straightedge to construct an equilateral triangle.

**15.** Complete the truth table.

| p | q | p → q | ~(p → q) | ~q | p ∧ ~q | ~(p → q) ↔ p ∧ ~q |
|---|---|-------|----------|-----|--------|-------------------|
| T | T |       |          |     |        |                   |
| T | F |       |          |     |        |                   |
| F | T |       |          |     |        |                   |
| F | F |       |          |     |        |                   |

**16.** In the figure below, $\overleftrightarrow{XY}$ and $\overleftrightarrow{VW}$ are both perpendicular to $\overleftrightarrow{MN}$. $\overline{PW}$ bisects $\angle YPQ$. What can you conclude about $\triangle PQW$? Explain.

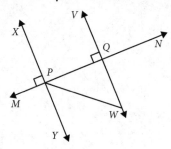

**17.** Create a truth table for $(p \vee \sim q) \rightarrow (\sim p \wedge q)$.

**18.** Define *parallel lines*.

**19.** Is the argument below valid? Explain.

*If a parallelogram is a rhombus, then the diagonals of the parallelogram are perpendicular to one another. The diagonals of parallelogram ABCD do not intersect to form right angles. Therefore, parallelogram ABCD is not a rhombus.*

**20.** What are the two key terms introduced in the definition of circle and what information about the circle do they provide?

## CHAPTER 3

# Geometry in the Coordinate Plane

The fundamental elements of geometry are points, lines, and planes. Examining points and lines on a coordinate plane adds important tools from algebra to geometric investigations.

## Points and Lines

The **Cartesian plane** is one method of organizing the points in a plane so that they can be referenced by ordered pairs of numbers. It takes the name *Cartesian* from the mathematician René Descartes. The system uses two number lines, one horizontal and one vertical, which intersect at the zero of each line. The horizontal number line is the **x-axis**, with positive numbers to the right and negatives to the left, while the vertical number line is the **y-axis**, with positives moving up and negatives down. Every point in the plane can be named by a pair of numbers, an **x-coordinate**, and a **y-coordinate**, in that order, that allow you to locate the point.

Lines are made up of points, but infinitely many points, so trying to list all the points on a line by their coordinates is impossible. Luckily, points on a line have coordinates that follow a pattern, and a line can be described by an equation that talks about the relationship of the *x*- and *y*-coordinates.

The **slope** of a line is a ratio that describes whether the line rises or falls, and how quickly. The numerator describes the vertical motion and the denominator the horizontal change. Lines with positive slopes rise and those with negative slopes fall. A horizontal line has a slope of 0. The larger the absolute value of a slope, the steeper the line will be. A line with a slope of $\frac{1}{2}$ would rise (left to right) slowly, while one with a slope of $-\frac{135}{7}$ would fall off quickly.

▶ If two points on the line are known, the slope can be found by the formula

$$m = \frac{y_2 - y_1}{x_2 - x_1}$$, where $(x_1, y_1)$ and $(x_2, y_2)$ are points on the line.

EXAMPLE

A line that passes through the points (–3, 5) and (2, –4) will have a slope of $m = \dfrac{-4 - 5}{2 - (-3)} = \dfrac{-9}{5}$.

▶ The **y-intercept** of a line is the $y$-coordinate of the point where the line intersects the $y$-axis.

▶ Every point on the $y$-axis has an $x$-coordinate of 0.

## EQUATIONS OF LINES

▶ **Slope-intercept form**  $y = mx + b$, where $m$ is the slope of the line and $b$ is the $y$-intercept. This form is convenient for quick graphing. The line $y = 2x - 5$ can be sketched by starting at –5 on the $y$-axis and counting up 2 and 1 to the right to place another point. Once two or more points are plotted, they can be connected into a line.

▶ **Point-slope form**  $y - y_1 = m(x - x_1)$, where $m$ is the slope of the line and $(x_1, y_1)$ is a point on the line. This form is preferred for finding the equation of the line when points are known. With two points, the slope can be calculated and the slope and one point inserted into this form. Point-slope form is often simplified down to slope-intercept form.

▶ A line that passes through the points (–3, 5) and (2, –4) will have a slope of $m = \dfrac{-4 - 5}{2 - (-3)} = \dfrac{-9}{5}$ as you saw above. Then you can choose one of the points to use in point-slope form.

$$y - y_1 = m(x - x_1)$$
$$y - y_1 = -\frac{9}{5}(x - x_1)$$
$$y - (-3) = -\frac{9}{5}(x - 5)$$
$$y + 3 = -\frac{9}{5}x + 9$$
$$y = -\frac{9}{5}x + 6$$

▶ **Standard (or general) form** $ax + by = c$, where $a$ and $b$ are **integer coefficients**, $c$ is an **integer constant**, and $a \geq 0$. This form is simply a rearrangement of slope-intercept form, and is useful for finding $x$- and $y$-intercepts for quick graphing.

▶ $$y = -\frac{9}{5}x + 6$$
$$\frac{9}{5}x + y = 6$$
$$5\left(\frac{9}{5}x + y\right) = 5(6)$$
$$9x + 5y = 30$$

▶ **Horizontal line** Because horizontal lines have a slope of 0, the equation of a horizontal line becomes $y = b$, where $b$ is a constant equal to the $y$-intercept. Every point on the horizontal line, no matter its $x$-coordinate, will have a $y$-coordinate equal to $b$.

▶ **Vertical line**   Attempting to calculate a slope for a vertical line results in a zero denominator, and so no slope can be found. The equation of a vertical line is $x = c$, where $c$ is a constant equal to the $x$-intercept. Every point on the vertical line has the same $x$-coordinate, regardless of its $y$-coordinate.

## EXERCISE 3.1

**Answer all questions. Use information about slope and the various forms of the equation of a line to respond to each exercise.**

**1.** Write the equation, in slope-intercept form, of the line shown below.

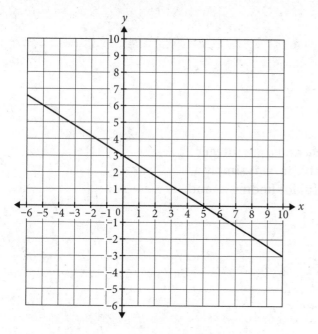

**2.** Find the slope of a line that passes through the points (–4, 5) and (5, –1).

**3.** Find the equation, in point-slope form, of a line that passes through the points (0, –5) and (6, 3).

**4.** Find the standard form of the equation of a line that passes through (7, –3) and (4, 8).

**5.** A line with a slope of $\frac{3}{5}$ passes through the points (1, –2) and (–3, $k$). Find the value of $k$.

**6.** Give the equations of the four lines shown in the figure below.

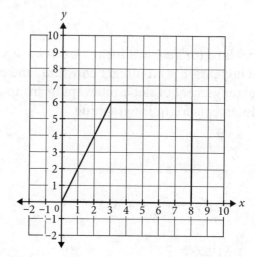

**7.** Is it possible for the line $3x - 5y = 35$ to pass through the point (107, 57)? Explain your reasoning.

# Parallels and Perpendiculars

In geometry, when we talk about parallel lines or perpendicular lines, the conversation usually turns toward angle measures, but coordinate geometry does not have a convenient way to measure an angle. Instead, the focus is on slopes.

▶ **Parallel** To demonstrate the two lines in the coordinate plane are parallel lines, show that they have the same slope. You can do this using the slope formula, or by putting the equation of the line in a form where the slope can easily be read.

▶ **Perpendicular** To show that two lines are perpendicular, that they meet at right angles, show that their slopes multiply to –1. Slopes that multiply to –1 are **negative reciprocals**. They have opposite signs, one positive and one negative, and are reciprocals of one another. Examples of negative reciprocals include $\frac{3}{5}$ and $-\frac{5}{3}$ or –7 and $\frac{1}{7}$.

## EXERCISE 3.2

**Answer all questions using the information about the slopes of parallel lines and of perpendicular lines. In questions 1 through 3, decide if the lines are parallel, perpendicular, or neither.**

**1.** $2x + 3y = 15$ and $3x - 2y = -14$

**2.** $2x - 3y = 3$ and $3x - 2y = -14$

**3.** $2x + 3y = 15$ and $2x + 3y = -9$

**4.** Find the equation of a line parallel to $y = \frac{1}{5}x - 2$ that passes through the point $(10, -5)$.

**5.** Find the equation of a line parallel to $2x - 7y = 14$ that passes through the point $(7, 2)$.

**6.** Find the equation of a line perpendicular to $y = -\frac{5}{8}x + 17$ that passes through the point $(-15, 4)$.

**7.** Find the equation of a line perpendicular to $5x + 3y = 0$ that passes through the point $(10, 7)$.

# Distance, Perimeter, and Area

To find perimeters or areas of figures in the coordinate plane, you first must find the lengths of line segments. If the line segment is horizontal or vertical, lengths are easily found by counting boxes on the grid, but for oblique, or slanted, lines, distance between two points can be found by using the distance formula: $d = \sqrt{(x_2 - x_1)^2 + (y_2 - y_1)^2}$, where $(x_1, y_1)$ and $(x_2, y_2)$ are the endpoints of the line segment. The distance formula is actually a variant of the Pythagorean theorem. Subtracting the $x$-coordinates gives you the length of one leg of a right triangle, and subtracting the $y$-coordinates gives the length of the other leg. The distance between the two points, or length of the segment, is the length of the hypotenuse.

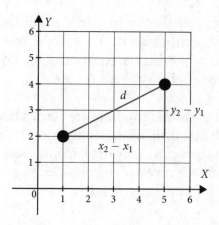

Once you can find the length of a segment, you can show that two segments are **congruent** (that is, have the same measurement), and you can find perimeters by adding the lengths of sides. If asked to find area, remember that the altitude, or height, must be perpendicular to the base. You may need to find that perpendicular line before finding the length of the altitude.

## EXERCISE 3.3

**Use the distance formula to construct a response to each exercise.**

1. Find the length of segment $\overline{AB}$ if $A$ is the point $(8, -6)$ and $B$ is $(12, -3)$.

2. What is the length of $\overline{XY}$ if $X$ is $(-3, 5)$ and $Y$ is $(9, 0)$.

3. Find the length of $\overline{RT}$ if $R$ is $(28, 10)$ and $T$ is $(16, 26)$.

4. The vertices of a triangle are $A$ $(3, 7)$, $B$ $(7, 4)$, and $C$ $(3, 2)$. Classify the triangle as equilateral, isosceles, or scalene.

   Questions 5 through 7 refer to quadrilateral $WXYZ$ with vertices $W$ $(0, 0)$, $X$ $(5, 12)$, $Y$ $(25, 12)$, and $Z$ $(20, 0)$.

5. Is $WXYZ$ a parallelogram? If so, show that both pairs of opposite sides are parallel. If not, explain.

6. Are opposite sides of $WXYZ$ congruent? Support your answer.

7. Find the area of $WXYZ$. Explain your reasoning.

# Dividing a Line Segment

Geometry looks at the equations of lines and relationships between lines or segments. Are they parallel, perpendicular, or neither? It looks at lengths of segments and calculations that can be done with those lengths, and it looks at ways that a line or segment might be divided. The most common question would be about bisecting a segment, and a line or segment bisects a segment if it passes through the midpoint of the segment.

The **midpoint** of the segment with **endpoints** $(x_1, y_1)$ and $(x_2, y_2)$ can be found by averaging the $x$-coordinates and averaging the $y$-coordinates.

$$M = \left( \frac{x_1 + x_2}{2}, \frac{y_1 + y_2}{2} \right)$$

EXAMPLE

The midpoint of a segment with endpoints (2, 3) and (5, 12)

is $M = \left( \dfrac{2 + 5}{2}, \dfrac{3 + 12}{2} \right) = \left( \dfrac{7}{2}, \dfrac{15}{2} \right)$.

Although there isn't a formula for dividing a segment into three (or four or five or ten) equal parts, there is a technique that only requires simple arithmetic. Subtract $x_2 - x_1$ and subtract $y_2 - y_1$. Divide each result by the number of parts you want.

EXAMPLE

Let's say you want to divide the line segment with endpoints (2, 3) and (5, 12) into three equal parts. $5 - 2 = 3$ and $12 - 3 = 9$. Divide 3 and 9 by 3 to get 1 and 3. The "low" endpoint is (2, 3), so add (2 + 1, 3 + 3) to get the point that divides the first section from the second, (3, 6). Repeat to find the end of the second section $(3 + 1, 6 + 3) = (4, 9)$. The third section is from (4, 9) to the endpoint (5, 12).

## EXERCISE 3.4

**Use the midpoint formula and the technique for dividing a segment into sections of equal length to work each exercise.**

1. Find the midpoint of segment $\overline{XY}$ if $X$ is (7, –1) and $Y$ is (–9, 3).

2. Find the midpoint of segment $\overline{JK}$ if $J$ is (6, 13) and $K$ is (–11, 2).

3. The midpoint of $\overline{AB}$ is M = (5, 3). If $A$ is (8, –7), what are the coordinates of $B$?

4. Triangle $\triangle ABC$ has coordinates $A$(5, 3), $B$(7, 11), and $C$(15, 3). Find the coordinates of $M$ and $N$, the midpoints of $\overline{AB}$ and $\overline{BC}$, respectively.

5. Find the midpoint $P$ of side $\overline{AC}$ in question 4.

6. Find the length of $\overline{MP}$ and the length of the side of $\triangle ABC$ to which $\overline{MP}$ is parallel.

7. In $\triangle ABC$ from question 4, find the points that divide $\overline{BC}$ into four sections of equal length.

8. Find the two points that divide segment $\overline{PQ}$ into equal thirds, if $P$ is the point (–4, 5) and $Q$ is (8, –10).

# Coordinate Proofs: Specific to General

At first glance, coordinate geometry is simple. If you're asked to show lines are parallel, show they have the same slope. To show line segments are congruent, find the length of each with the distance formula and see if they come out the same. Does one line segment bisect another? Find the midpoint and see if that's where they cross.

But to *prove* something, to show that it's not just an accident of this particular figure, you have to show that it is true not just in this one specific figure, but in every figure that matches this description. It might be possible to show that one particular quadrilateral *JKLM* is a square, but that doesn't prove that every quadrilateral is a square. For proof, you need to move from the specific to the general.

When you look at a specific figure on the coordinate plane, you have the coordinates of key points; you have numbers. You can do arithmetic to find lengths, slopes, and midpoints. To construct a proof, you want to draw your figure on a coordinate plane, but instead of specific numbers, you want to use letters as placeholders or parameters for the coordinates.

Let's suppose you want to show that the diagonals of a parallelogram bisect each other, by showing that both diagonals have the same midpoint. Draw a parallelogram on the coordinate plane. Start with one vertex at the origin (0, 0). Let one side of the parallelogram sit on the positive *x*-axis. The other end of that segment is a point on the *x*-axis, so its *y*-coordinate is 0, but what is the *x*-coordinate? We don't want to give a specific number, so call it (*a*, 0).

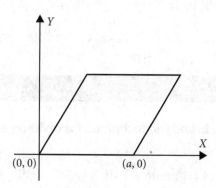

Draw the opposite side parallel to the *x*-axis. That's easy because it's horizontal. But how long is it? Well, it's *a* units long, but what is that? Don't worry too much for the drawing; just make it look about the same size. But remember, the left and right edges won't be vertical, so shift the top over a bit and connect top to bottom with slanted lines that look roughly parallel. In other words, make it look like a parallelogram.

We don't know much about how to label the upper left corner. Call the coordinates $(b, c)$. The last vertex is a little easier. The $x$-coordinate will be $a$ units more than $b$, and the $y$-coordinate is the same as $c$. Label it $(b + a, c)$. Then draw the diagonals.

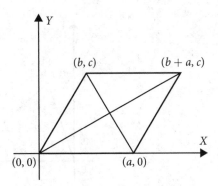

You want to show that the point where the two diagonals cross is the midpoint of both diagonals. Use the midpoint formula to find the two midpoints, just as you would if you had numbers. It may look a little messy, but it's the same process.

Longer diagonal: $M = \left( \dfrac{0 + b + a}{2}, \dfrac{0 + c}{2} \right) = \left( \dfrac{b + a}{2}, \dfrac{c}{2} \right)$

Shorter diagonal: $M = \left( \dfrac{b + a}{2}, \dfrac{c + 0}{2} \right) = \left( \dfrac{b + a}{2}, \dfrac{c}{2} \right)$

Both diagonals have the same midpoint, so they bisect one another. And because you showed that for any parallelogram, you have proved it is the case for the diagonals of any parallelogram.

> **Tips:**
> - Place one vertex at the origin, if possible.
> - Let segments fall on axes, if possible.
> - Don't assume every coordinate needs a new letter. Think about relationships to find ways to reuse labels.
> - If in doubt about what to do, ask yourself what you would do if you had numbers.

## EXERCISE 3.5

**For each exercise, draw a diagram if one is not provided, and prove the statement, showing all calculations.**

1. Prove that the diagonals of a rectangle are congruent.

2. Prove that the diagonals of a square are perpendicular.

**3.** Use the figure below to prove that *ABCD* is a trapezoid.

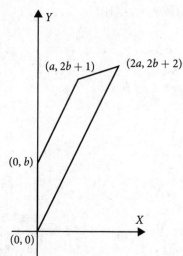

**4.** Use the figure above to prove that the line segment that connects the midpoints of the nonparallel sides is parallel to the bases.

**5.** Given $\triangle ABC$ with $\overline{BD} \perp \overline{AC}$ in the diagram below. Prove that if $\overline{BD}$ bisects $\overline{AC}$, then $\triangle ABC$ is isosceles.

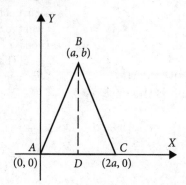

**6.** Using $\triangle XYZ$ below, find the midpoint of each of the three sides, and draw the line segments that connect each vertex to the midpoint of the opposite side.

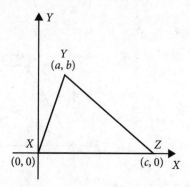

**7.** Using the line segment drawn from the origin in question 6, find the points that divide the segment into three equal parts.

**8.** Prove that in an isosceles right triangle, the line segment from the right angle vertex to the midpoint of the hypotenuse is perpendicular to the hypotenuse.

## CHAPTER 4

# Transformations

A **transformation** is an operation that moves or changes an object in a clearly defined fashion. **Rigid transformations**, like reflection, translation, and rotation, preserve the size and shape of the object, but change its location or its orientation, or both. Rigid transformations produce images that are congruent to the original, or pre-image. *Dilations*, which are covered in a later chapter, stretch or shrink the object, and produce images that are similar to the pre-image.

## Reflections

Reflections are the fundamental transformation upon which translation and rotation are built. A **reflection** mirrors an object across a specified line called the **reflecting line**. Reflections (and other transformations) can be seen as functions that work on points rather than numbers. The domain is the set of points that make up the original object or pre-image. Each point of the pre-image is sent to an image point across the reflecting line.

Given a pre-image and a reflecting line, you can find the image by constructing a line through a point on the pre-image, perpendicular to the reflecting line.

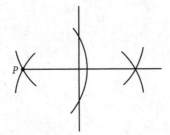

Measure the distance from the pre-image point to the reflecting line, and mark the image point on the new line, an equal distance on the opposite side of the reflecting line.

> The image of a point *P* is a point *P′* positioned on the other side of the reflecting line so that the reflecting line is the perpendicular bisector of $\overline{PP'}$.

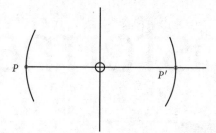

Any figure you might want to reflect is made up of many, even infinitely many, points. If you had to move each point by construction, the task would be endless. Fortunately, it is only necessary to reflect a few key points, and then connect them.

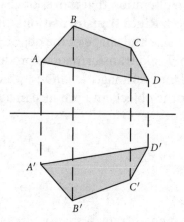

Reflecting by construction can be a tedious process, but devices and technology exist to create reflections quickly and easily. In between those extremes is the option of creating reflections on the coordinate plane, where perpendicular lines are easier to spot and there is a formula for distance. Common reflections, over axes or lines like $y = x$, can be summarized by these rules:

▶ Reflection over the *x*-axis changes the sign of the *y*-coordinate of each point.

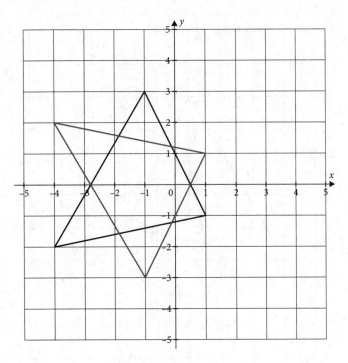

▶ Reflection over the *y*-axis changes the sign of the *x*-coordinate.

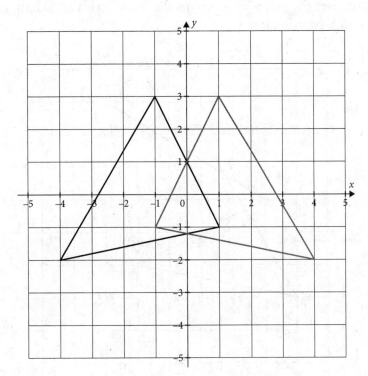

▶ Reflection over the line $y = x$ swaps the $x$- and $y$-coordinates.

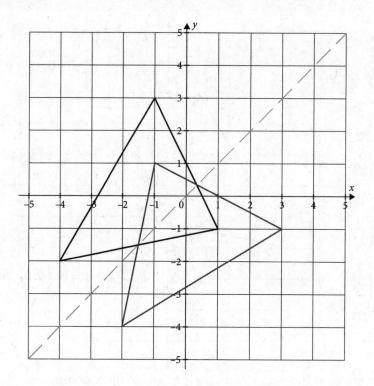

▶ Reflection over the line $y = -x$ swaps the coordinates and changes the signs of both coordinates.

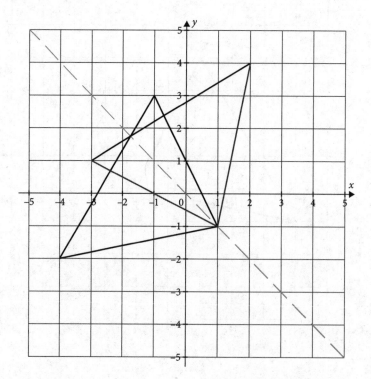

## EXERCISE 4.1

**For exercises 1 and 2, use a compass and a straightedge to construct the reflection image.**

1. Use a compass and a straightedge to reflect point $P$ over line $\overleftrightarrow{AB}$.

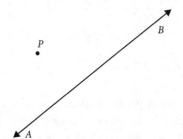

2. Use a compass and a straightedge to reflect $\triangle ABC$ over line $\overleftrightarrow{XY}$.

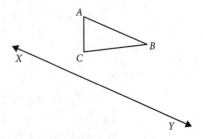

**In questions 3 through 5, find the coordinates of the vertices of the image and sketch.**

3. The vertices of $\triangle ABC$ are $A(4, 5)$, $B(-3, 2)$, and $C(2, -3)$. Reflect $\triangle ABC$ over the $y$-axis.

4. The vertices of $\triangle RST$ are $R(1, 4)$, $S(3, 1)$, and $T(2, -4)$. Reflect $\triangle RST$ over the $x$-axis.

5. The vertices of rectangle $WXYZ$ are $W(-1, 1)$, $X(7, 1)$, $Y(7, -2)$, and $Z(-1, -2)$. Reflect $WXYZ$ over the line $y = x$.

**Questions 6 through 8 ask you to make observations about the reflections you sketched.**

6. Use the distance formula

$$d = \sqrt{(x_2 - x_1)^2 + (y_2 - y_1)^2}$$ and your results in exercise 3 to show that $AB = A'B'$. (Reflection preserves length.)

7. In exercise 4, $\triangle RST$ was named by moving in a clockwise direction around the vertices. Is $\triangle R'S'T'$ also oriented clockwise or has reflection changed the orientation?

8. In exercise 5, you reflected a rectangle across $y = x$. Use $WXYZ$ and $W'X'Y'Z'$ to show that angle measurement has been preserved; that is, that right angles in the pre-image result in right angles in the image.

# Translations

The transformation referred to as **translation** is usually imagined as a simple slide, but is, in fact, created by two reflections, in sequence, across parallel reflecting lines. The distance between the pre-image and the reflecting lines, and the order in which they are used, determine the distance the object is translated. If the reflecting lines are vertical, the object is translated horizontally. Horizontal reflecting lines result in vertical translation.

Oblique reflecting lines will cause an oblique translation, perpendicular to the reflecting lines, but the same result can be obtained by a horizontal translation followed by a vertical translation.

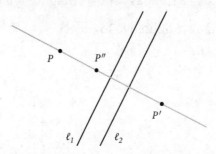

Reflection preserves length and angle measure, so translation will as well. In the previous exercises, you saw that reflection reverses orientation. Since translation is two reflections, orientation is reversed and then reversed again, so in the end, translation preserves orientation.

When viewed on the coordinate plane, translation can be seen as a function that sends each point $(x, y)$ of the pre-image to a point whose $x$-coordinate or $y$-coordinate, or both, are increased or decreased by a fixed amount.

EXAMPLE

> A translation $T$ moves a point of the form $(x, y)$ to a point $(x + h, y + k)$. Symbolically, $T(x, y) \rightarrow (x + h, y + k)$, where $h$ and $k$ are constants. If $h$ is positive, the movement is to the right; if $h$ is negative, the point moves left. A negative value of $k$ signifies a downward move, while a positive $k$ denotes an upward move.

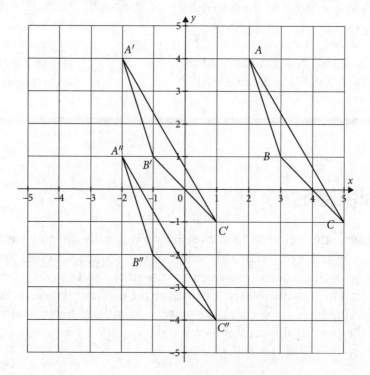

## EXERCISE 4.2

**These exercises focus on translations as the composition of two reflections. Use what you know about reflection as well as the material about translations in this section to answer each question.**

1. Reflection is a rigid transformation because it produces an image that is congruent to the pre-image. Explain why translation also produces an image identical in size and shape to the pre-image.

2. Use a compass and a straightedge to translate point $P$ by reflecting first over $\overleftrightarrow{AB}$ and then over $\overleftrightarrow{CD}$.

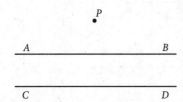

3. Use a compass and a straightedge to translate point $P$ by reflecting first over $\overleftrightarrow{CD}$ and then over $\overleftrightarrow{AB}$. Is the result the same as in question 2? Explain.

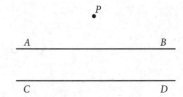

**Questions 4 through 7 focus on translations in the coordinate plane. Work with coordinates as much as possible, but sketch the translation if it helps you.**

4. Find the image of the point $(-3, 1)$ under the translation $T(x, y) \rightarrow (x + 5, y - 2)$.

5. Find the image of the point $(2, 0)$ under the translation $T(x, y) \rightarrow (x, y - 4)$.

6. $\triangle ABC$ has vertices $A(2, -3)$, $B(-1, 7)$, and $C(4, 3)$. Give the vertices of $\triangle A'B'C'$, the image of $\triangle ABC$ under the translation $T(x, y) \rightarrow (x + 3, y - 2)$.

7. Quadrilateral $ABCD$ has vertices $A(-1,1)$, $B(7,1)$, $C(7, -2)$, and $D(-1, -2)$. Give the vertices of $A'B'C'D'$, the image of $ABCD$ under $(x, y) \rightarrow (x + 4, y - 1)$.

8. Give an example of a sequence of translations that would carry a polygon onto itself.

# Rotations

Like translation, **rotation** is a rigid transformation that is created by two reflections in sequence, but unlike translation, which is reflections over parallel lines, a rotation is created by reflecting over intersecting lines. The effect is to rotate the pre-image about the point at which the two reflecting lines intersect.

The rotation is described by its size, measured in degrees, and its direction, clockwise or counterclockwise. The size of the rotation is twice the number of degrees in the smaller angle between reflecting lines.

▶ If the reflecting lines form angles of 36° and 144°, the object will rotate 2(36°) = 72° around the point at which the lines intersect. If the rotation is counterclockwise, it would be indicated as positive 72°, and if it is clockwise, it would be −72°.

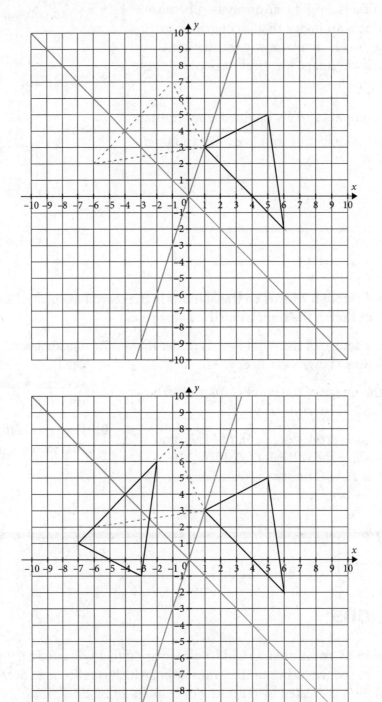

When rotation is drawn on the coordinate plane, the rotation is generally centered at the origin, and it's possible to describe the effect of the rotation by describing the changes to the coordinates of key points. This is generally limited, however, to multiples of 90°, because the calculation for other angles becomes complicated:

$R_{90}$ A rotation of 90° about the origin moves the point from $(x, y)$ to $(-y, x)$.

$R_{180}$ A rotation of 180° about the origin moves the point from $(x, y)$ to $(-x, -y)$.

$R_{270}$ A rotation of 270° about the origin moves the point from $(x, y)$ to $(y, -x)$.

$R_{-90}$ A rotation of −90° about the origin moves the point from $(x, y)$ to $(y, -x)$.

$R_{-180}$ A rotation of −180° about the origin moves the point from $(x, y)$ to $(-x, -y)$.

$R_{-270}$ A rotation of −270° about the origin moves the point from $(x, y)$ to $(-y, x)$.

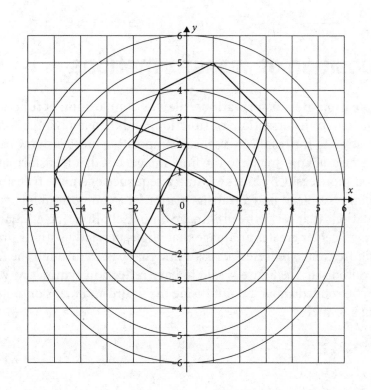

## EXERCISE 4.3

**Respond to each exercise about rotation using the method of reflections or the measurement method as appropriate.**

1. Explain why direction is insignificant in a rotation of 180°.

2. $A(2, -3)$, $B(-1, 7)$, and $C(4, 3)$ are the vertices of $\triangle ABC$. Find the image $\triangle A'B'C'$ of $\triangle ABC$ under a rotation of 270° counterclockwise about the origin.

3. What other rotation about the origin would carry the $\triangle ABC$ (in question 2) on to $\triangle A'B'C'$?

4. If $X(2, -3)$, $Y(-1, 7)$, and $Z(4, 3)$ are the vertices of $\triangle XYZ$, find the image $\triangle X'Y'Z'$ after a rotation of 180° about the origin.

5. Compare the rotation in question 4 to reflecting $\triangle XYZ$ over the $y$-axis.

6. Rotate a quadrilateral with vertices $A(-1,1)$, $B(7,1)$, $C(7, -2)$, and $D(-1, -2)$ $-270°$ about the origin.

7. Use what you know about rotation and translation to explain how you could determine the result of rotating a figure $-90°$ about the point $P(4, -2)$.

# Composition of Transformations

Keep in mind that the order in which the transformations are applied may produce different results.

A translation followed by a reflection may produce a different image than the reflection followed by the translation.

The **composition of transformations**, like the composition of functions, refers to applying one transformation (or function) and then applying a second transformation to the result. In a sense, composition creates a new function that carries the original pre-image to the final image, but that transformation can be understood more easily by breaking into a sequence of simple transformations.

It is often possible to find a transformation or a composition of transformations that carry the pre-image onto itself. Rotating a square 90° about its own center creates an image that aligns perfectly with its pre-image. When a figure can be reflected across a line that passes through the figure and map onto itself, the figure is said to have **reflection symmetry**. When a figure can be rotated about a point inside the figure and maps onto itself, it has **rotational symmetry**.

## EXERCISE 4.4

**These exercises combine different transformations. Choose the method you find most useful for drawing the transformation. Be careful to execute the transformations in the correct order.**

1. What is the image of $(4, -1)$ under a counterclockwise rotation of $90°$ followed by a translation 5 units left and 4 units up?

2. Rectangle $WXYZ$ has vertices $W(1, 3)$, $X(1, 5)$, $Y(5, 5)$, and $Z(5, 3)$. Find the image of $WXYZ$ under a rotation of $180°$ about the origin, followed by a reflection over the $x$-axis.

3. Find the image of $WXYZ$ in question 2 under a reflection over the $x$-axis, followed by a rotation of $180°$ about the origin. Does changing the order of the transformations alter the result?

4. Over what line(s) could you reflect the regular pentagon to carry it onto itself?

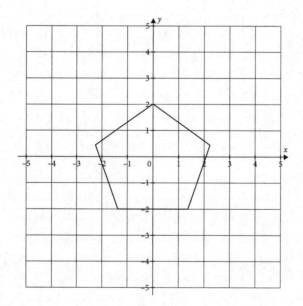

5. What rotation(s) about the origin would carry the regular hexagon onto itself?

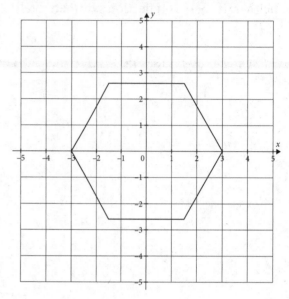

6. Describe a sequence of transformations that carry $\triangle ABC$ onto $\triangle A'B'C'$. Explain your thinking.

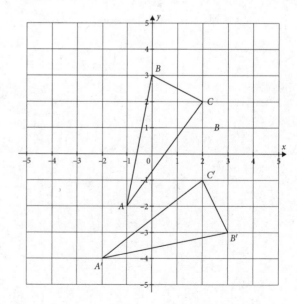

**7.** An image has reflection symmetry if a reflection over a line carries the image onto itself. It has rotational symmetry if a rotation about a point within the image carries it onto itself. For each letter of the alphabet shown below, decide if the letter has reflection symmetry, rotational symmetry, both, or neither. If there is reflection symmetry, tell whether the reflection line is horizontal or vertical. If there is rotation symmetry, through what angle does the letter rotate?

ABCDEFG
HIJKLM
NOPQRST
UVWXYZ

# Coordinate Geometry and Transformations

The review questions in this section cover the ideas from Chapters 3 and 4 and allow you to see if you can apply all the ideas of coordinate geometry correctly. You'll also be able to check your understanding of the four basic transformations. Answer all the questions, and try to express your thinking as clearly as you can.

**1.** Sketch the graph of $y = -2x + 5$ on the axes below.

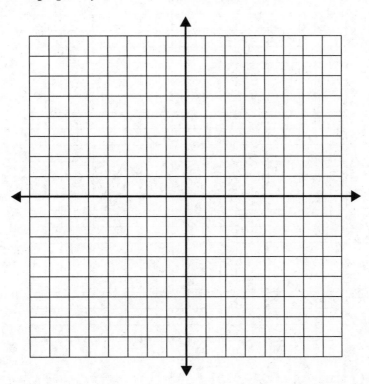

2.  Find the slope of a line that passes through the points $(-3, -1)$ and $(5, 5)$.

3.  What is the slope of the line represented by the equation $2x + 3y = 2(x - 1) + 5$?

4.  Find the equation of a line that passes through $(2, -7)$ and $(-4, 5)$. Give the equation in point-slope form.

5.  Give the equation in standard form of a line that passes through the points $(2, 1)$ and $(-3, -1)$.

6.  Determine whether the equations below represent lines that are parallel, perpendicular, or neither.

$$2x - 7y = -7$$
$$7x + 2y = -10$$

7.  Find the equation of a line parallel to $5y = x + 17$ that passes through $(3, 2)$.

8.  Find the midpoint of $\overline{AB}$ if $A$ is $(6, -5)$ and $B$ is $(-4, 3)$.

9.  Find the equation of the perpendicular bisector of the segment that connects $(-7, 3)$ to $(3, -7)$.

10. Give a coordinate proof that the diagonals of a rectangle are congruent.

11. Draw the image of the triangle under a reflection over the line $y = x$.

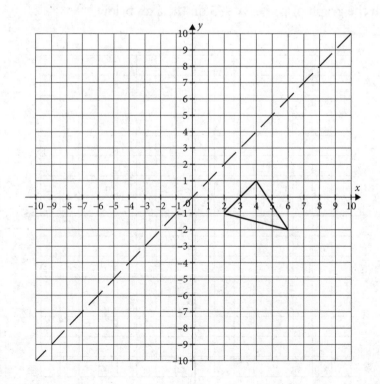

**12.** Draw the image of the triangle under a rotation of 90° counterclockwise about the origin.

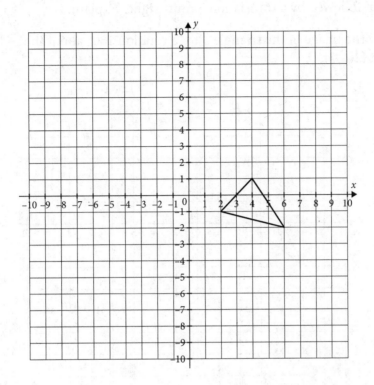

**13.** Describe the translation that carries $\triangle PQR$ onto $\triangle P'Q'R'$.

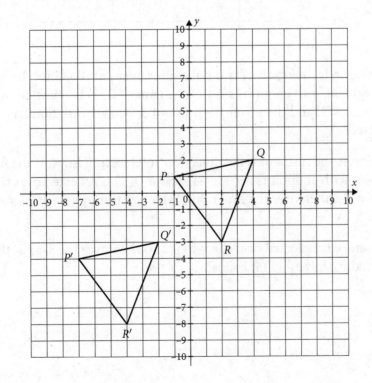

14. Will a translation 3 units right followed by a 90° clockwise rotation about the origin produce the same image as a 90° clockwise rotation about the origin followed by a translation 3 units right? Explain.

15. Draw the image of the triangle after a rotation 180° about the point (2, −3).

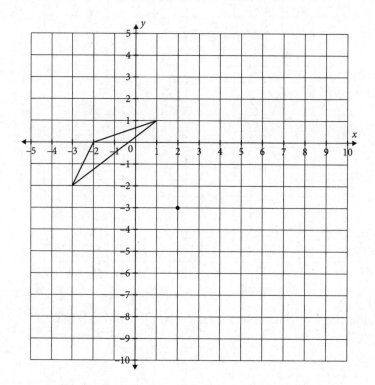

16. Translation is the result of two reflections over parallel lines. If the line segment with endpoints A (−1, −1) and B (1, 1) is reflected over the line $x = 3$, and then over the line $x = 5$, what translation has occurred?

17. Rotation is the result of reflections over intersecting lines. If the line segment with endpoints A (−1, −1) and B (1, 1) is reflected over the line $y = −x$ then over the line $y = x$, what are the coordinates of $A'$ and $B'$?

18. If an image is reflected over the $y$-axis and then over the $x$-axis, through how many degrees will it rotate?

**19.** When a polygon has line symmetry, there is at least one line over which a reflection will carry the polygon onto itself. Draw these lines of symmetry for the polygon shown below.

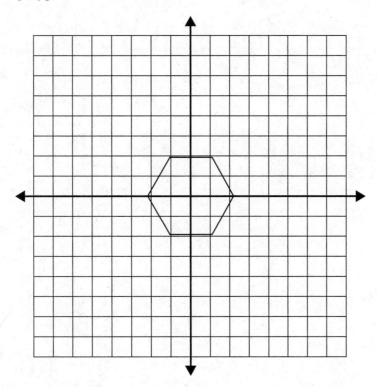

**20.** The polygon above also has rotational symmetry. Describe the rotation(s) that carry it onto itself.

## CHAPTER 5

# Triangles

Geometry is the study of many different shapes, but the most familiar is probably the triangle. The simplest closed figure with straight sides, the triangle may be the figure with the most information to discover.

## Triangle Vocabulary

Triangles are key to much of the work in geometry. They provide a convenient way of breaking down and investigating other figures.

Triangles can be classified by sides or by angles. A triangle with three sides of equal length is **equilateral**. If two sides of a triangle are congruent, the triangle is **isosceles**. If each of the sides is a different length, the triangle is **scalene**.

A triangle that contains a right angle is a **right triangle**, and a triangle that contains an angle of more than 90° is an **obtuse triangle**. If all three angles are less than 90°, the triangle is an **acute triangle**. A triangle with three angles of equal size is **equiangular**.

In an isosceles triangle, the sides of equal length are called **legs**. The angle formed by the two congruent sides is called the **vertex angle**. The side that is not congruent to the other two is called the **base**, and an angle formed when the base meets a leg is a **base angle**.

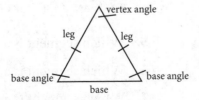

In a right triangle, the two sides that meet to form the right angle are called *legs* and the third side, opposite the right angle, is the **hypotenuse**.

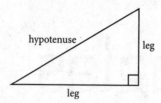

A line segment drawn from the vertex of a triangle to the midpoint of the opposite side is called a **median**. A line segment from the vertex perpendicular to the opposite side is an **altitude**. In an obtuse triangle, an altitude may fall outside the triangle. In that case, a side is extended and the altitude meets that extension at a right angle. A **midsegment** of a triangle is a line segment that connects the midpoints of two sides of a triangle.

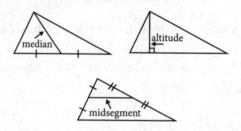

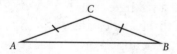

## EXERCISE 5.1

**Answer all of these questions about the vocabulary associated with triangles. Use the figure below for questions 1 to 3.**

Use the figure below and the following information for exercises 5 to 7.

1. Which side of isosceles triangle △ABC is the base?

2. Is the vertex angle of △ABC larger or smaller than one of the base angles?

3. Classify △ABC based on its angles.

4. Drawing a midsegment in a triangle creates a smaller triangle. Explain why the new triangle will have the same classification, by sides and angles, as the original triangle.

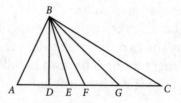

$\overline{BD} \perp \overline{AC}$, $\angle ABE \cong \angle CBE$, and $\overline{AF} \cong \overline{FC}$.

5. Which segment is an altitude of △ABC?

6. Which segment is a median of △ABC?

7. Which segment is an angle bisector of △ABC?

# Triangle Sum Theorem

One of the first pieces of information we're given about triangles is that the three angles of a triangle have measurements that add to 180°. This **Triangle Sum theorem**, like any theorem, should be proved, not just accepted. The proof begins with a construction, but it is not necessary to actually perform the construction. You know it is possible to construct a line parallel to a given line, through a given point.

**EXAMPLE**

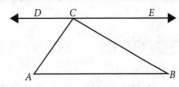

▶ Suppose we construct line $\overleftrightarrow{DE}$ through point $C$, so that $\overleftrightarrow{DE} \parallel \overline{AB}$. The other two sides of the triangle can act as transversals, and we can say $\angle DCA \cong \angle CAB$ and $\angle ECB \cong \angle CBA$, because when parallel lines are cut by a transversal, alternate interior angles are congruent. We can see that $\angle DCA$, $\angle ACB$, and $\angle ECB$ together form straight angle $\angle DCE$, which measures 180°. We can write $m\angle DCA + m\angle ACB + m\angle ECB = m\angle DCE$ and then substitute:

$$m\angle DCA + m\angle ACB + m\angle ECB = m\angle DCE$$

$$m\angle CAB + m\angle ACB + m\angle CBA = 180°$$

$$m\angle A + m\angle C + m\angle B = 180°$$

▶ Because the three angles of a triangle always add to 180°, it's possible to find the measure of the third angle whenever two angle measurements are known. It also tells us that a triangle cannot contain more than one right angle, or more than one obtuse angle. Equiangular triangles contain three 60° angles, and the two acute angles of a right triangle add to 90°.

▶ **Triangle Sum theorem**   The measures of the three interior angles of any triangle always total 180°.

**EXERCISE 5.2**

**Use the Triangle Sum theorem to answer each of these questions.**

**1.** In $\triangle RST$, $m\angle R = (3x - 7)°$, $m\angle S = (5x + 2)°$, and $m\angle T = (x + 5)°$. Find the value of $x$ and the measure of each angle, in degrees.

**2.** Classify $\triangle RST$ from question 1 according to its angles.

3. In $\triangle ABC$, m$\angle A$ is five times m$\angle B$, and m$\angle C$ is the sum of m$\angle A$ and m$\angle B$. Find the measure of each angle, in degrees.

4. Classify $\triangle ABC$ from question 3 according to its angles.

5. $\triangle XYZ$ is a right triangle with $\overline{XY} \perp \overline{YZ}$. m$\angle X = (5a - 16)°$ and m$\angle Z = (70 - 2a)°$. Find the measure of each angle.

6. In quadrilateral $ABCD$, diagonal $\overline{AC}$ is drawn. Show that the measures of the four angles in the quadrilateral total 360°.

7. Show that the five angles of a pentagon total 540°.

8. If $n$ is the number of sides of a polygon, explain why the total number of degrees in the $n$ interior angles of the polygon will total $180°(n - 2)$.

# Base Angle Theorem

If two sides of a triangle are congruent, the triangle appears to have a symmetry that suggests the angles opposite those congruent sides have equal degree measures. In other words, the base angles of an isosceles triangle are congruent.

EXAMPLE

▶ To prove this, start with an isosceles triangle in which $\overline{AB} \cong \overline{BC}$. Construct the bisector of the vertex angle of isosceles $\triangle ABC$. (You don't actually have to stop and do the construction. You know you can bisect an angle.) The angle bisector will intersect $\overline{AC}$ at point $D$, but don't make any assumptions about $D$. Just reflect $\triangle ABC$ over $\overline{BD}$. If you think the image looks a lot like $\triangle ABC$, that's good. Because the reflecting line is inside $\triangle ABC$, the pre-image and image are on top of one another, and that's hard to keep track of, so let's draw them separately.

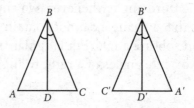

▶ Reflection is a rigid transformation, so it will preserve length and angle measure. $\overline{BD}$ is the reflecting line so points on $\overline{BD}$ are unchanged, which means $\overline{BD} \cong \overline{B'D'}$. Because $\overline{BD}$ bisected $\angle ABC$, $\angle ABD \cong \angle CBD$, so when the triangle is reflected over $\overline{BD}$, $\angle ABD$ is carried onto $\angle CBD$ and $\angle CBD$ onto $\angle ABD$. $\overline{AB} \cong \overline{BC}$ because these are the congruent sides of the isosceles triangle. $B$ is $B'$ because $B$ is on the reflecting line, so if $\overline{AB} \cong \overline{BC}$, then $A$ must map onto $C$ and $C$ onto $A$ to preserve the lengths. If $A$ and $C$ are carried onto one another, the $\overline{AC}$ is carried onto $\overline{CA}$. The reflection carries the vertices and sides of $\angle BAC$ onto those of $\angle BCA$, and preserves angle measure, so $\angle BAC \cong \angle BCA$.

▶ **Base Angle theorem** If two sides of a triangle are congruent, the angles opposite those sides are congruent.

The converse of the Base Angle theorem says that if two angles of a triangle are congruent, the sides opposite those angles are congruent. The proof of the converse is similar to the proof of the theorem, and you can complete it in the exercises.

▶ **Base Angle Converse theorem** If two angles of a triangle are congruent, the sides opposite those angles are congruent.

## EXERCISE 5.3

**Use the Base Angle theorem and its converse (and any theorems you learned previously) to answer these questions.**

1. Prove that if two angles of a triangle are congruent, then the sides opposite those angles are congruent.

2. If the vertex angle of isosceles triangle $\triangle PQR$ measures 94°, what is the measure of each base angle?

3. If one base angle of isosceles triangle $\triangle ABC$ measures 37°, what is the measure of the vertex angle?

4. Given equilateral triangle $\triangle RST$, use the Base Angle theorem to show that the equilateral triangle is also equiangular.

5. $\triangle ABC$ is isosceles with $\overline{AB} \cong \overline{BC}$ and $\triangle DBE$ is isosceles with $\overline{BD} \cong \overline{BE}$.

Prove that m$\angle BAC$ = m$\angle EBC$ + m$\angle E$.

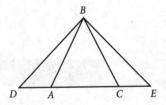

6. Use the results of question 5 to prove $\angle DBA \cong \angle EBC$.

7. In isosceles triangle $\triangle XYZ$, m$\angle X$ = $(3x - 1)°$, m$\angle Y$ = $(2x + 3)°$. Is it possible to identify which are the base angles and which is the vertex angle? Explain your reasoning.

## Concurrence

When three or more lines meet at a single point, the lines are said to be **concurrent**. In any triangle, there are several sets of lines that can be drawn or constructed and shown to be concurrent. The points of concurrence are referred to as *centers* of the triangle.

If all three medians of a triangle are drawn, they will meet at a point inside the triangle. This point is called the **centroid** of the triangle. The location of the centroid is twice as far from the vertex as from the opposite side.

If each of the angles of the triangle is bisected, the three angle bisectors meet at a point inside the circle called the **incenter**. The incenter is equidistant from the sides of the triangle. If the perpendicular bisector of each of the sides is constructed, the three perpendicular bisectors will meet at a point called the **circumcenter.** The circumcenter may be a point inside the triangle, outside the triangle, or on a side of the triangle, and is equidistant from the vertices of the triangle.

The three altitudes of a triangle meet at a point called the **orthocenter,** which may fall inside, outside, or on the triangle.

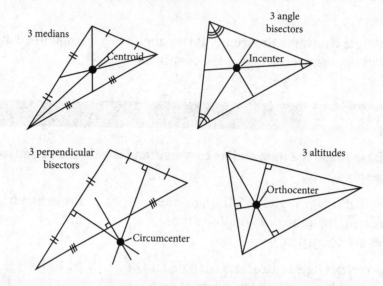

## EXERCISE 5.4

These exercises focus on the concurrences covered in this section. Use facts about the centroid, incenter, circumcenter, and orthocenter to answer each question.

1. In the figure below, $P$ is the centroid of $\triangle ABC$. $PD = 6$ cm, $PE = 4$ cm, and $AC = 18$ cm. Find the perimeter of $\triangle AEP$.

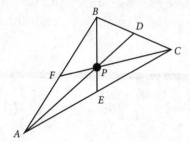

2. $P$ is the incenter of $\triangle XYZ$. $PR = 4$ cm, $XQ = 11$ cm, $ZS = 12$ cm, and $XZ = 14$ cm. Find the perimeter of $\triangle XPZ$.

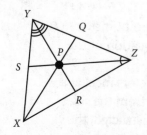

3. If $\triangle ABC$ is a right triangle, where will its orthocenter fall? Explain your reasoning.

4. Is it possible for the incenter and the circumcenter to be the same point? Explain your reasoning.

5. $P$ is the incenter of equilateral triangle $\triangle ABC$. Prove that $\overline{AD} \cong \overline{EC}$.

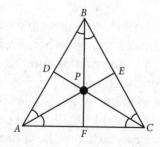

6. In the figure below, $P$ is the centroid of $\triangle ABC$. $CP = x + 7$, $PF = x + 1$, $BE = 4x + 1$, and $BC = 5x - 1$. Find the perimeter of $\triangle BPC$.

7. Use a compass and a straightedge to construct the circumcenter of $\triangle RST$.

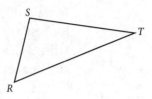

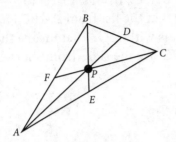

# Inequalities in One Triangle

In any triangle, the longest side of the triangle is opposite the largest angle, and the shortest side opposite the smallest angle. Conversely, the smallest angle lies opposite the shortest side and the largest angle opposite the longest side. This relationship is why we know that the hypotenuse of a right triangle is always the longest side, because it sits opposite the right angle, the largest angle in the triangle.

▶ The longest side of a triangle is opposite the largest angle, and the shortest side is opposite the smallest angle.

▶ The largest angle of a triangle is opposite the longest side, and the smallest angle is opposite the shortest side.

**EXAMPLE**

▶ In the figure below, $\triangle ABC$ shares a side with $\triangle ACD$. The length of each side is marked on the diagram. Can you put the angles in order from smallest to largest?

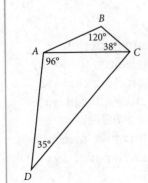

▶ Start in △ABC, find the third angle (22°), and put the angles in order from smallest to largest: 22°, 38°, and 120°. The sides opposite these angles will rank from shortest to longest: $\overline{BC}$, $\overline{AB}$, and $\overline{AC}$. Move to △ACD and find the third angle there (48°). Put the angles in order from smallest to largest: 35°, 48°, and 96°. The opposite sides, from shortest to longest are $\overline{AC}$, $\overline{AD}$, and $\overline{DC}$. Notice that the list from △ABC ends with $\overline{AC}$ and the list from △ACD begins with $\overline{AC}$. That means the lists can be merged: $\overline{BC}$, $\overline{AB}$, $\overline{AC}$, $\overline{AD}$, $\overline{DC}$ are the sides from shortest to longest.

Right triangles have sides that fit a clearly defined relationship, called the *Pythagorean theorem*, which we'll look at later. Unfortunately, other triangles don't have such a consistent rule. There is one thing that is always true about the sides of a triangle, however. The length of a side of a triangle is always less than the sum of the lengths of the other two sides, and more than their difference. You can have a triangle with sides of 21 inches, 28 inches, and 30 inches. 28 − 21 < 30 < 28+21, 30 − 21 < 28 < 30 + 21, and 30 − 28 < 21 < 30 + 28. But you cannot have a triangle with sides of 4, 7, and 11 inches because 4 + 7 = 11.

▶ **Triangle Inequality theorem**   The length of the third side of a triangle is less than the sum of the lengths of the other two sides but more than their difference.

If one side of the triangle is extended beyond a vertex, the extended side and the side adjacent to it form an **exterior angle**. The exterior angle is supplementary to the interior angle adjacent to it. With a little algebra, you can prove that an exterior angle of a triangle is equal to the sum of the two remote interior angles.

EXAMPLE

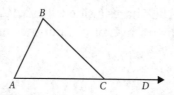

▶ In △ABC, extend side $\overline{AC}$ through vertex C to D, forming exterior angle ∠BCD. m∠ACB + m∠BCD = 180°. By the Triangle Sum theorem, m∠BAC + m∠ABC + m∠ACB = 180°. The two sums are both equal to 180°, so m∠ACB + m∠BCD = m∠BAC + m∠ABC + m∠ACB. Subtract m∠ACB from both sides and m∠BCD = m∠BAC + m∠ABC. The measure of the exterior angle, ∠BCD, is equal to the sum of the measures of the two interior angles that are not adjacent to the exterior angle.

▶ **Exterior Angle theorem**   The measure of an exterior angle of a triangle is equal to the sum of the measures of the two nonadjacent angles.

Once you know that m∠BCD = m∠BAC + m∠ABC, it is clear than one of the interior angles is less than the exterior angle. So m∠BCD > m∠BAC and m∠BCD > m∠ABC.

▶ **Exterior Angle Inequality theorem**   The measure of an exterior angle of a triangle is larger than the measure of either of the two nonadjacent angles.

## EXERCISE 5.5

**Use the Triangle Inequality theorem to answer questions about side lengths and the Exterior Angle theorem and Exterior Angle Inequality theorem to answer questions about angle measures.**

1. If a triangle is to be constructed with sides of 18 feet and 25 feet, what is the range of possible lengths for the third side?

2. Side $\overline{AC}$ of △ABC is extended through C to D, forming exterior angle ∠BCD. If m∠A = 53° and m∠BCD = 122°, what is the m∠B?

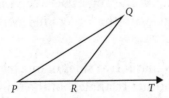

3. In the figure below, m∠PRQ = (7x + 2)° and m∠QRT = (6x − 4)°. If m∠QPR = 43°, what is the measure of ∠PQR?

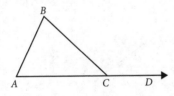

4. In △ABC below, side $\overline{AC}$ is extended through C to create exterior angle ∠BCD, and $\overline{AE}$ bisects ∠BAC. Prove that m∠B = m∠AEC − m∠EAC.

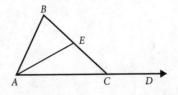

5. List the segments in the figure below in order from shortest to longest.

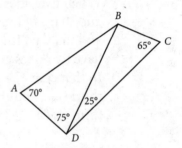

6. List the angles in the figure below in order from largest to smallest.

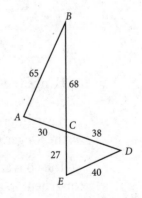

7. In the figure below, sides of △DBC and △EBA are extended to create exterior angles ∠BAD and ∠BCE. If ∠BAD ≅ ∠BCE, prove △ABC is isosceles.

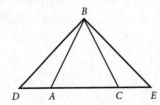

**8.** In $\triangle ABC$, one side is extended at each vertex, creating one exterior angle at each vertex. Show that the sum of the measures of these exterior angles is 360°.

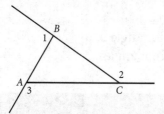

**9.** $\triangle PQR$ is a right triangle with $\overline{PQ} \perp \overline{QR}$. If $m\angle R > m\angle P$, which is the shortest side of $\triangle PQR$?

# Inequalities in Two Triangles

When two sides of a triangle and the angle between them are set, the size and shape of the triangle is fixed. You can connect the end of the given sides and that's it. The **Hinge theorem** talks about a situation in which two triangles have two sides congruent to two sides, but the included angles are different. In that case, the triangle that has the smaller included angle has the shorter third side, and the triangle with the larger included angle has the longer third side.

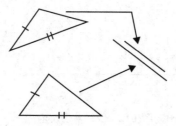

▶ **Hinge theorem** If two sides of one triangle are congruent to two corresponding sides of another triangle, but the angles between the sides have different measurements, the triangle with the larger angle will have the longer third side.

The converse of the Hinge theorem says that if two sides of one triangle are congruent to the corresponding sides of another triangle, but the third sides are not congruent, then the triangle with the longer third side has the larger included angle.

▶ **Hinge converse theorem** If two sides of one triangle are congruent to two corresponding sides of another triangle, but the third sides have different lengths, the triangle with the longer third side will have the larger included angle.

## EXERCISE 5.6

**Use the Hinge theorem and its converse to answer each of these questions.**

1. In the figure below, $AB = BC$. If m∠$ABD$ = 43° and m∠$CBD$ = 46°, how do the lengths of $AD$ and $CD$ compare?

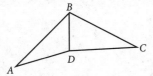

2. Two isosceles triangles △$PQR$ and △$RQT$ share side $QR$. If $PQ = QR = QT$ and $PR > RT$, how do the measures of ∠$PQR$ and ∠$RQT$ compare?

3. In the figure below, $BD$ is a median and m∠$ADB$ ≤ m∠$BDC$. Compare the lengths of $AB$ and $BC$. Explain your thinking.

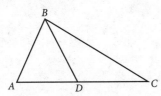

4. The basic kite shape is two isosceles triangles with the same base, and their vertices in opposite directions. In the top triangle, the altitude is small and in the bottom one, the altitude is longer. Which triangle will have the larger vertex angle? Explain your reasoning.

5. In the figure below, $AB = BC$. If you want to determine whether $AD$ is longer than $CD$, what other information do you need?

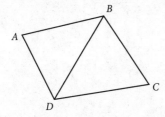

6. △$WXY$ is isosceles with $\overline{WX} \cong \overline{XY}$ and point Z is on $\overline{WY}$ such that $WZ < ZY$. Explain why $XZ$ cannot be the bisector of ∠$WXY$.

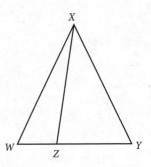

7. In the figure below, $\overline{AD} \cong \overline{CD}$, m∠$ADB$ = 35°, and m∠$CDB$ = 47°. Prove that $AB < BC$.

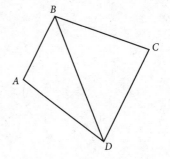

## CHAPTER 6

# Right Triangles

Right triangles occur so commonly in our experience that they deserve a bit of special mention. In addition to their prevalence, they warrant attention because they open the door to a branch of mathematics known as trigonometry. That will come up in a later chapter, but for now, let's take a closer look at this category of triangles.

## Right Triangles

The definition of a **right triangle** simply says that it is a triangle containing one right angle. Because the Triangle Sum theorem tells us that the measures of the three angles of a triangle sum to 180°, it is not possible to have more than one right angle in a triangle. Two right angles, 90° each, would total 180°, leaving no possible third angle.

The presence of a right angle in a triangle leads to several interesting properties:

▶ The two sides that meet to form the right angle are called the legs.

▶ The legs are perpendicular to one another, and therefore each is an altitude of the triangle.

▶ When one leg is taken as the altitude, the other leg is the base.

▶ The area of a right triangle is one-half the product of the legs of the legs.

▶ The third side of the right triangle, the hypotenuse, is opposite the right angle. The side opposite the largest angle is the longest side, so the hypotenuse is always the longest side of the right triangle.

▶ The two acute angles of a right triangle are always complementary. The right angle measures 90° so the other two must total to another 90°.

▶ The shorter leg is opposite the smaller acute angle, and the longer leg is opposite the larger acute angle.

## EXERCISE 6.1

**Use the information about sides and angles of right triangles to answer these questions.**

1. If $m\angle A = 17°$ and $m\angle B = 74°$, is $\triangle ABC$ a right triangle? Explain.

2. In right triangle $\triangle XYZ$, with $\overline{XY} \perp \overline{YZ}$, $m\angle X = 47°$. Which is the shortest side of $\triangle XYZ$?

3. The acute angles of right triangle $\triangle ABC$ are $\angle A$ and $\angle B$. If $m\angle A = 7x - 11$ and $m\angle B = 6x + 10$, find the measure of each acute angle.

4. List the sides of $\triangle ABC$ in exercise 3 in order from shortest to longest.

5. The area of a right triangle is 480 cm² and the hypotenuse measures 52 cm. Find the length of the altitude to the hypotenuse to the nearest tenth of a centimeter.

6. Right triangle $\triangle RST$ has side lengths $RS = (2x + 4)$ cm, $ST = \left(\frac{1}{2}x\right)$ cm, and $RT = (3x - 13)$ cm. If the perimeter of the triangle is 90 cm, which angle is the right angle?

7. A right triangle has sides whose lengths are integers, and an area of 84 square centimeters. The altitude to the hypotenuse measures exactly 6.72 cm. Find the length of the hypotenuse and the possible lengths of the legs. Explain your reasoning.

# The Pythagorean Theorem

One of the best known properties of right triangles is the relationship among their sides. The **Pythagorean theorem** says that the square of the length of the hypotenuse is equal to the sum of the squares of the other two sides. This rule is usually summarized by labeling the legs of the right triangle as $a$ and $b$ and the hypotenuse as $c$. If $a$ and $b$ are the lengths of the legs of a right triangle and $c$ is the length of the hypotenuse, then $a^2 + b^2 = c^2$.

There are several different ways to prove the Pythagorean theorem, and famous ones originate in different cultures and different times. One constructs squares on each side of the right triangle and then shows that the two smaller squares can be rearranged to fit exactly in the largest square. Another relies on a bit of algebra. Four copies of the right triangle are arranged as shown below. They form a large square, which measures $a + b$ on each side. The area of that large square is $(a + b)^2 = a^2 + 2ab + b^2$. The triangles each have an area of $\frac{1}{2}ab$ and there are four of them, so the area shaded below is $4\left(\frac{1}{2}ab\right) = 2ab$.

Taking $2ab$ away from $a^2 + 2ab + b^2$ leaves $a^2 + b^2$ for the center white square, but you can also see that the area of that square is $c^2$. $a^2 + b^2 = c^2$.

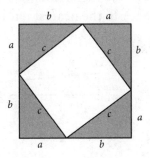

The Pythagorean theorem can be used to find the length of any side of a right triangle, if the other two side lengths are known. If the lengths of the legs are 5 cm and 12 cm, the length of the hypotenuse, squared, equals $5^2 + 12^2 = 25 + 144 = 169$. If $c^2 = 169$, then $c = \sqrt{169} = 13$ cm. If the hypotenuse is 20 cm, and one leg is 8 cm, $8^2 + b^2 = 20^2$. Square and then subtract to find $b^2 = 400 - 64 = 336$. In this case again, $b = \sqrt{336}$ but 336 is not a perfect square, so either leave the value of $b$ in simplest radical form $b = \sqrt{336} = \sqrt{16 \cdot 21} = 4\sqrt{21}$ or use a calculator and appropriate rounding to get $b = \sqrt{336} \approx 18.33$ cm.

When square roots are involved, perfect squares whose roots are integers are not common. When a set of three integers that fit the $a^2 + b^2 = c^2$ rule turn up, we call them a **Pythagorean triple**. Examples of Pythagorean triples include the set 3, 4, 5 and the set 5, 12, 13. There are others, but also note that multiples of Pythagorean triples are also Pythagorean triples. Because 3, 4, 5 is a triple, so are 6, 8, 10 and 45, 60, 75 and 300, 400, 500.

## EXERCISE 6.2

**Apply the Pythagorean theorem to answer each of these questions.**

1. Find the length of the hypotenuse of a right triangle with legs that measure 8 cm and 15 cm.

2. The hypotenuse of a right triangle measures 100 inches, and one leg measures 28 inches. Find the length of the other leg.

3. Alex planted a young tree and wanted to stabilize it by running wires from the tree to stakes in the ground. One wire will be attached to the tree at a spot 7 feet above the ground. If it must be pulled tightly to an anchor in the ground 4 feet from the base of the tree, how long, to the nearest tenth of a foot, should the wire be?

4. In right triangle $\triangle ABC$, leg $\overline{BC}$ measures 10 cm, and hypotenuse $\overline{AC}$ is 2 cm longer than leg $\overline{AB}$. Find the length of leg $\overline{AB}$.

**5.** If the area of right triangle $\triangle RST$ is $A = (8x)$ cm$^2$ and its perimeter is $(2x + 18)$ cm, find $x =$ the length of leg $\overline{RS}$. Explain your reasoning.

**6.** For each of the triangles below, the lengths of two sides are given. Find the third side and tell whether they form a Pythagorean triple.

   a) $a = 7, b = 9, c = ?$

   b) $a = ?, b = 21, c = 29$

   c) $a = 9, b = ?, c = 41$

**7.** The recommended design for a wheel chair ramp says that for every 1 inch rise, it should run 12 inches (1 foot). If a ramp needs to rise 18 inches and observe the 1:12 rule, how long, to the nearest tenth of an inch, is the ramp itself?

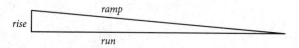

# Converse of the Pythagorean Theorem

The Pythagorean theorem begins with the hypothesis that the triangle is a right triangle, and then tells how the lengths of the sides are related. The converse of the theorem begins with the relationship among the side lengths, and tells the type of triangle: right, acute, or obtuse.

   If $a$, $b$, and $c$ are the lengths of the sides of a triangle, and $c$ is the longest side, then the following are true:

▶ If $a^2 + b^2 = c^2$, then the triangle is a right triangle.

▶ If $a^2 + b^2 < c^2$, then the triangle is an obtuse triangle.

▶ If $a^2 + b^2 > c^2$, then the triangle is an acute triangle.

**EXAMPLE**

▶ If a triangle has sides of 8, 11, and 15 cm, let $c$ be the longest side, and then $c^2 = 15^2 = 225$. Square each of the other side lengths, and add them: $a^2 + b^2 = 8^2 + 11^2 = 64 + 121 = 185$. Compare 185 to 225, and see that $a^2 + b^2 < c^2$. A triangle with sides of 8, 11, and 15 cm will be obtuse. If sides of 8 and 11 were placed to form a right angle, the 15-cm side would be too long to form a hypotenuse. The sides of 8 and 11 would need to be rotated into an obtuse angle to make room for the 15-inch side.

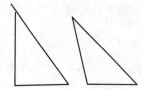

**Use the Triangle Inequality theorem, the Pythagorean theorem, and the converse of the Pythagorean theorem to answer these questions.**

In questions 1 through 4, the lengths of three sides of a triangle are given. Is it possible to make a triangle with the side lengths given? If so, is the triangle acute, right, or obtuse? Show your reasoning.

1.  5, 8, and 11

2.  23, 17, and 41

3.  65, 52, and 39

4.  12, 17, and 19

The Triangle Inequality theorem gives the largest and smallest possible lengths for the third side of a triangle, if two sides are known. The converse of the Pythagorean theorem gives additional information.

5.  Suppose you want to construct an acute triangle with sides of lengths 5 cm and 7 cm. What does the Triangle Inequality theorem tell you about possible lengths for the third side? If you want the third side to be the longest side, what lengths are possible for the third side?

6.  Suppose you want to construct an obtuse triangle with sides of lengths 8 cm and 12 cm. What does the Triangle Inequality theorem tell you about possible lengths for the third side? If you want the third side to be the longest side, what lengths are possible for the third side?

7.  If you want to construct an acute triangle with sides of 7 and 11, what lengths are possible for the third side?

# Special Right Triangles

The Pythagorean theorem provides a method for finding the third side of any right triangle if two sides are known, but there are two families of right triangles, called **special right triangles**, for which that work is almost unnecessary.

Isosceles right triangles, or 45°–45°–90° right triangles, have two legs of the same length. When you use the Pythagorean theorem to find the hypotenuse, you notice that, no matter what the length of the leg is, $c = \sqrt{a^2 + a^2} = \sqrt{2a^2} = a\sqrt{2}$. The length of the hypotenuse of a 45°–45°–90° right triangle is always the length of a leg times the square root of 2.

EXAMPLE

▶ If the leg is 17 cm, the hypotenuse is $17\sqrt{2}$ cm. If the leg is 528 feet, the hypotenuse is $528\sqrt{2}$ feet.

▶ You can work in the other direction as well. If the hypotenuse of a 45°–45°–90° right triangle is $43\sqrt{2}$ meters, the leg is 43 meters. If the hypotenuse is 26 inches, the leg is $\dfrac{26}{\sqrt{2}}$, which simplifies to

$$\frac{26}{\sqrt{2}} = \frac{26\sqrt{2}}{2} = 13\sqrt{2} \text{ inches.}$$

The other group of right triangles that have easy-to-remember side lengths are the 30°–60°–90° right triangles. Triangles of this type are created by drawing an altitude in an equilateral triangle. You can see that the shortest side, opposite the 30° angle, is half as long as the hypotenuse.

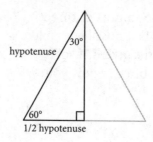

When you apply the Pythagorean theorem, the third side is:

$$b = \sqrt{c^2 - a^2}$$

$$= \sqrt{(hypotenuse)^2 - \left(\tfrac{1}{2}hypotenuse\right)^2}$$

$$= \sqrt{\tfrac{3}{4}(hypotenuse)^2}$$

$$= \tfrac{1}{2}hypotenuse\sqrt{3}$$

The legs of a 30°–60°–90° right triangle are half the hypotenuse (opposite the 30° angle) and half the hypotenuse times the square root of three (opposite the 60° angle).

EXAMPLE

If you know that the shortest side is 12 cm, the other leg is $12\sqrt{3}$ cm and the hypotenuse is 24 cm. If you know the hypotenuse is 28 inches, the legs are 14 inches and $14\sqrt{3}$ inches. If you know the longer leg is

39 feet, the shorter leg is $\dfrac{39}{\sqrt{3}} = \dfrac{39\sqrt{3}}{3} = 13\sqrt{3}$ feet and the hypotenuse is $26\sqrt{3}$ feet.

In a 45°–45°–90° right triangle, the legs are $x$
and $x$ and the hypotenuse is $x\sqrt{2}$.

$\downarrow$

side-side-side radical 2

In a 30°–60°–90° right triangle, the hypotenuse
is $h$ and the legs are $\frac{1}{2}h$ and $\frac{1}{2}h\sqrt{3}$.

$\downarrow$

hypotenuse-half the hypotenuse-half the hypotenuse radical 3

## EXERCISE 6.4

**All the questions below involve either 30–60°–90° triangles or 45°–45°–90° right triangles. Use properties of special right triangles to work these exercises.**

1. If the leg of an isosceles right triangle measures 64 cm, how long is the hypotenuse?

2. If the hypotenuse of a 30°–60°–90° right triangle is 18 meters, how long are the legs?

3. A square has a side of 19 inches. How long is a diagonal of the square?

4. If the shortest side of a 30°–60°–90° right triangle measures 34 cm, how long is the other leg?

5. If the hypotenuse of an isosceles right triangle measures 22 inches, how long is the leg?

6. If the hypotenuse of a 30°–60°–90° right triangle is 36 cm, how long is the longer leg?

7. If the area of an isosceles right triangle is 98 square centimeters, how long is the hypotenuse?

# Triangles, Right Triangles

These practice questions bring together the concepts covered in Chapters 5 and 6 and focus on information about triangles, especially the important right triangle. Answer all the questions, and try to express your thinking as clearly as you can.

**1.** Classify each triangle by sides and angles.

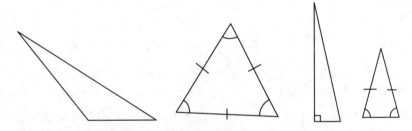

**2.** Find the measure of each angle if m∠A = 2x + 3, m∠ACB = 5x − 2, m∠B = 89 − x, and m∠BCD = 8x − 13.

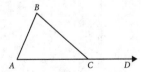

**3.** In △MNO, m∠M = 58° and m∠N = 63°. Order the sides of △MNO from shortest to longest.

**4.** In △RST, side $\overline{RS}$ is extended to create exterior angle ∠TSV. m∠TSV = 5(x − 1), m∠R = 4x − 5, and m∠RTS = x. Find the measure of ∠TSR.

5. $\triangle ABC$ is an isosceles triangle with $\overline{AB} \cong \overline{BC}$. m$\angle A = 2x + 10$, m$\angle B = x + 10$, and m$\angle C = 3x - 20$. Find the measure of the base angles.

6. In $\triangle XYZ$, medians $\overline{XA}$, $\overline{YB}$, and $\overline{ZC}$ intersect at point $P$, the centroid. If $XP = 28$ cm, $PB = 16$ cm, and $YZ = 54$ cm, find the perimeter of $\triangle PAY$.

7. In the figure below, $\overline{OA} \cong \overline{OC}$ and $\overline{OB} \cong \overline{OD}$. $AB < CD$. What conclusion can you draw about the measures of angles in the figure? Explain.

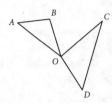

8. Jasmine has a small garden that is roughly triangular. Two sides are already bordered by benches, and Jasmine wants to purchase fencing to border the third side. She measured the two sides with benches as 3 feet and 5 feet long. What are the minimum and maximum lengths of fencing she might need to border the third side?

9. In the figure below, $\angle XZW \cong \angle YZW$ and $\angle ZXY \cong \angle ZYX$. Prove $\triangle XWY$ is isosceles.

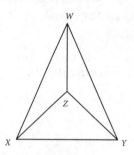

10. In the figure below, $\triangle ACE$ and $\triangle BDG$ are both isosceles triangles with $\overline{AC} \cong \overline{CE}$ and $\overline{BG} \cong \overline{GD}$. If $\overline{AE} \parallel \overline{BD}$, prove that $\angle HBA \cong \angle FDE$.

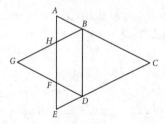

11. A right triangle has legs that measure 36 inches and 77 inches. How long is the hypotenuse of the triangle?

12. A certain right triangle has one leg that measures 13 cm, and another that is 1 cm shorter than the hypotenuse. Find the perimeter of the right triangle.

13. The hypotenuse of a right triangle measures 25 meters. Both legs have integer lengths, which can be represented by $x - 8$ and $x + 9$. Find the length of the shorter leg.

14. A triangle has legs of 8 inches, 11 inches, and 14 inches. Is the triangle acute, right, or obtuse?

15. If an isosceles triangle has legs twice as long as the base, is the triangle acute, right, or obtuse?

16. If an isosceles triangle has a third side that is equal to the length of a leg times the square root of two, is the triangle acute, right, or obtuse?

17. Jason and Brianna were trying to construct a model of a local park that is shaped like a triangle, with sides as shown in the figure below. There is a path from one end of the 119-meter side, perpendicular to the 153-meter side. The path meets the 153-meter side at a right angle at a point 100 meters from the other end of the 119-meter side. To the nearest tenth of a meter, how long is the path?

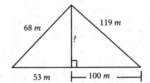

18. As Jason and Brianna tried to construct an accurate model of the triangular park, they wondered whether the vertex from which the path originated was a right angle or not. Explain how they can determine whether it is right, acute, or obtuse.

19. In right triangle $\triangle ABC$, the altitude from the right angle $\angle B$ intersects hypotenuse $\overline{AC}$ at $D$. If $AD = 8$ cm and $BD = 9$ cm, how long is $DC$?

20. In right triangle $\triangle RST$, with right angle $\angle S$, altitude $\overline{SU}$ divides hypotenuse $\overline{RT}$ into two sections. $UT = 21$ cm and $ST = 29$ cm. Find the length of hypotenuse $\overline{RT}$ to the nearest centimeter.

## CHAPTER 7

# Congruence

When you work with numbers, the principal relationship is equality. When you move to geometry, you come to understand that when the lengths of two segments (numbers) or the degree measures of two angles (numbers) are equal, then the segments or the angles are **congruent**. Numbers are equal, but objects are congruent. In this chapter, you'll look at what it means to say that two more complex objects, like triangles or other polygons, are congruent.

## Congruence

Two polygons are congruent if it is possible to establish a correspondence between them so that corresponding sides are congruent and corresponding angles are congruent. In simpler terms, two polygons are congruent if they are the same shape and the same size. Size is determined by the lengths of the sides, so the longest side of one will be congruent to the longest side of the other, and so on for the other sides. To have the same shape, the angles must match up from one triangle to the other.

If there is a rigid transformation (or a composition of rigid transformations) that carries one polygon onto the other, then the sides that align will have the same length, and the angles that align will have the same measure. The corresponding sides and corresponding angles are congruent.

EXAMPLE

In the figure below, reflecting $\triangle RST$ over the line $x = 4$, carries it onto $\triangle CBA$. $\overline{RS} \cong \overline{CB}$, $\overline{ST} \cong \overline{BA}$, and $\overline{RT} \cong \overline{CA}$. In addition, $\angle R \cong \angle C$, $\angle S \cong \angle B$, and $\angle T \cong \angle A$.

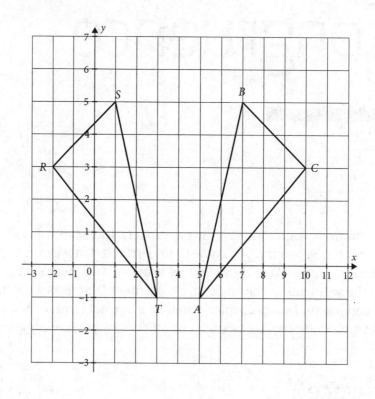

## EXERCISE 7.1

**Use the definition of congruence and the concept of corresponding parts to answer the following questions.**

1. If $\triangle CAT \cong \triangle DOG$, list the corresponding angles and corresponding sides.

2. In the example in this lesson, why would it not be correct to say $\triangle RST \cong \triangle ABC$? If you wished to say that the triangle on the left was congruent to $\triangle ABC$, how should you write that congruence statement?

3. If you have two triangles and know that $\angle R \cong \angle G$, $\angle S \cong \angle H$, $\angle T \cong \angle I$, and that $\overline{RS} \cong \overline{GH}$, $\overline{ST} \cong \overline{HI}$, and $\overline{RT} \cong \overline{GI}$, can you say the triangles are congruent? If so, write the correct congruence statement. If not, explain why.

4. Describe a rigid transformation that carries $\square ABCD$ onto $\square WXYZ$. What is the correct

congruence statement for the correspondence under that transformation?

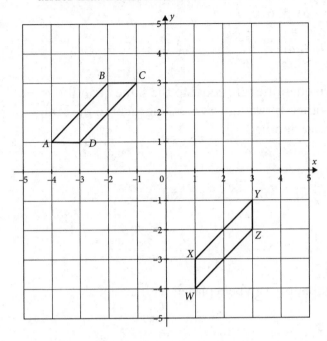

5. If ▱*ABCD*, on the grid to the left, is rotated 90° clockwise around the origin, it is carried onto ▱*PQRS*. Draw ▱*PQRS* on the grid to the left, and list the corresponding angles and corresponding sides.

6. Explain why any triangle is congruent to itself. Write the congruence statement.

7. Explain why an isosceles triangle is congruent to itself and why there are two different possible correspondences.

# Tips for Proofs

Before you explore the idea of proving two triangles congruent, it may help to take a minute to review some things you already know, so that you know how to cite them in a proof. Whenever you're asked to prove a statement, the question you should always be asking yourself, in case someone else challenges you, is how do you know your statement is true? Your answer to that question needs to be that the information was given in the question, or the theorem or postulate that you've previously learned that applies, or some fundamental principle of arithmetic or algebra that you're using. Here are some examples.

▶ **Reflexive property** Any quantity equals itself; any figure is congruent to itself.

**EXAMPLE**

$14 = 14$; $\triangle ABC \cong \triangle ABC$; $a = a$

▶ **Symmetric property** An equality relationship or a congruence relationship can be read from left to right or right to left.

EXAMPLE

> If $5 + 4 = 9$, then $9 = 5 + 4$. If $\triangle ABC \cong \triangle XYZ$, then $\triangle XYZ \cong \triangle ABC$. If $0 = x^2 + 5x + 6$, then $x^2 + 5x + 6 = 0$

▶ **Transitive property**    If two quantities are equal to the same quantity, then they are equal to each other. If two figures are congruent to the same figure, they are congruent to each other.

EXAMPLE

> If $a = b$ and $b = c$, then $a = c$. If $\triangle ABC \cong \triangle XYZ$ and $\triangle XYZ \cong \triangle RST$, then $\triangle ABC \cong \triangle RST$.

▶ **Substitution property**    A quantity can be replaced with an equal quantity.

EXAMPLE

> If $a = b$ and $b + c = d$, then $a + c = d$

This property goes by several different names depending on what you're working on: segment addition property, angle addition property, the whole is equal to the sum of its parts. But the version above can be abbreviated as SPEW, which is silly, but fun.

▶ **Sum of the Parts Equal the Whole**    If a point $P$ lies on a line segment, the distance from one endpoint to $P$ plus the distance from $P$ to the other endpoint is exactly equal to the length of the line segment. If a ray is drawn from the vertex of an angle and falls between the sides of the angle, it creates two smaller angles and the measures of those two angles add to exactly the measure of the original angle.

EXAMPLE

> If $C$ is a point on line segment $\overline{AB}$, then $AC + CB = AB$. If ray $\overrightarrow{AB}$ is in the interior of $\angle CAD$, then $m\angle CAB + m\angle BAD = m\angle CAD$.

▶ **Addition, Subtraction, Multiplication, and Division Properties of Equality**    These are the principles you learned when solving equations. If you add (or subtract) the same quantity or equal quantities on each side of an equation, the result is an equivalent equation. If you multiply both sides of the equation by the same quantity, the resulting equation has the same solution. If you divide both sides by a nonzero number, the result is an equivalent equation.

**EXAMPLE**

> If $3x - 7 = 17$, then $3x - 7 + 7 = 17 + 7$. If $m\angle A + m\angle B = m\angle B + m\angle C$, then $m\angle A + m\angle B - m\angle B = m\angle B - m\angle B + m\angle C$.

The other tips for working on proofs have to do with diagrams:

▶ If you aren't given a diagram, draw your own.

▶ Mark the diagram with tick marks to show what segments are congruent and arcs to show what angles are congruent. Invent your own marks if it helps.

▶ If figures overlap in the diagram, it may help to draw them separately.

# SSS (Side-Side-Side) and SAS (Side-Angle-Side)

The definition of congruence describes the relationships between all pairs of corresponding parts, three pairs of sides and three pairs of angles, but fortunately, it is not necessary to check that every pair of corresponding parts are congruent every time you want to prove triangles congruent. In particular, there are shortcuts for deciding if two triangles are congruent. If, for example, you can determine that the three pairs of corresponding sides largest, middle, and smallest—are congruent, it's not possible for the triangles to have corresponding angles that don't match. Those three side lengths can only be assembled into one shape. As a result, if you can prove that three pairs of corresponding sides are congruent, you can conclude that the triangles are congruent.

▶ **SSS Congruence**  If the three sides of one triangle are each congruent to the corresponding sides of another triangle, then the triangles are congruent and the corresponding angles are congruent as well.

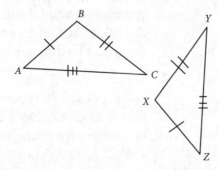

Another powerful shortcut for proving that triangles are congruent is also built on the idea that once these three pieces are set, the remaining parts of the triangles must match as well. If the lengths of two pairs of sides are known to be congruent, and the angle included between those sides is the same in both triangles, the triangles will be congruent. Once the two sides and the angle they form are fixed, the only triangle that can be created is formed by connecting the two sides.

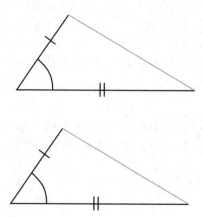

▶ **SAS Congruence** If two sides of one triangle and the angle included between them are each congruent to the corresponding parts of another triangle, then the triangles are congruent.

## EXERCISE 7.2

**Use SSS or SAS to prove triangles congruent in these exercises. It may help you to mark an A or an S next to each pair of corresponding parts you prove congruent, so you can see when you have achieved your goal.**

1. $\triangle ABC$ is an isosceles triangle with $\overline{AC} \cong \overline{BC}$. Segment $\overline{CM}$ connects vertex $C$ to $M$, the midpoint of $\overline{AB}$. Prove that $\triangle ACM \cong \triangle BCM$.

2. Chris claims that any two isosceles triangles with bases of equal length are congruent. Do you agree or disagree with Chris's contention? If you agree, construct a proof. If you disagree, give a counterexample (an example showing the statement is not true).

3. In the figure below, $\overline{MN}$ and $\overline{PQ}$ intersect at $O$ and bisect one another. Prove $\triangle MPO \cong \triangle NQO$.

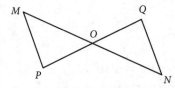

4. $\triangle ABC$ is an isosceles triangle with $\overline{AC} \cong \overline{BC}$. Segment $\overline{CD}$ bisects $\angle ACB$. Prove that $\triangle ACD \cong \triangle BCD$. Explain why $\overline{CD}$ is actually the same segment as $\overline{CM}$ described in question 1.

5. $ABCD$ is a rhombus, a parallelogram in which $\overline{AB} \cong \overline{BC} \cong \overline{CD} \cong \overline{DA}$. Diagonal $\overline{BD}$ is drawn. Prove $\triangle ABD \cong \triangle CDB$.

6. If question 5 began with "$ABCD$ is a rectangle" would it still be possible to prove $\triangle ABD \cong \triangle CDB$? In what ways would the argument be different?

7. Suppose $\triangle XYZ$ is not an isosceles triangle and suppose $\overline{YM}$ connects vertex $Y$ to $M$, the midpoint of $\overline{XZ}$. Does $\overline{YM}$ divide $\triangle XYZ$ into two congruent triangles? Does $\overline{YM}$ bisect $\angle XYZ$? Explain your reasoning.

# ASA (Angle-Side-Angle) and AAS (Angle-Angle-Side)

The other two commonly used shortcuts for proving two triangles congruent are variants of one another. The first says that if you can prove that two angles of one triangle are congruent to the corresponding angles of another triangle, and that the sides that connect the angles are also congruent, then the triangles will be congruent. If you set the length of a side and place angles whose sizes are fixed at each end of that side, the triangle is completed when the sides of those two angles intersect. There is no other way to shape a triangle.

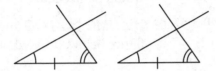

▶ **ASA Congruence**    If two angles and the included side of one triangle are congruent to the corresponding parts of a second triangle, then the triangles are congruent.

The other shortcut is a variant of ASA, based on the fact that we know the three angles of a triangle always add to 180°. If you know that a side of one triangle is congruent to a side of another triangle, and you know that two angles of the first triangle are congruent to the corresponding angles of the other triangle, but the parts aren't positioned correctly for you to apply ASA, it's still possible to prove the triangles congruent. In the figure below, for example, the sides that are known to be congruent are not included between the angles whose measures are known. But in each triangle, you could show that the third angle measures $180° - (22° + 108°) = 50°$. That would allow you to argue those third angles are congruent as well. Then you could prove the triangles congruent by ASA. Instead of having to show the third angles congruent every time it comes up, we allow one another the AAS shortcut.

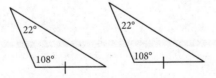

▶ **AAS Congruence**    If two angles and a side not included between them are congruent to the corresponding parts of another triangle, then the triangles are congruent.

## EXERCISE 7.3

**Use ASA or AAS to prove triangles congruent in these exercises.**

1. In the figure below, $\angle RST \cong \angle RQP$, and $\overline{SR} \cong \overline{QR}$. Prove $\triangle PQR \cong \triangle TSR$.

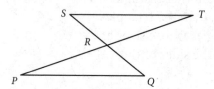

2. In the figure above, if the given information was that $\overline{ST} \cong \overline{PQ}$, rather than $\overline{SR} \cong \overline{QR}$, what angles must be congruent in order to prove the triangles congruent by ASA?

3. In the figure below, $\angle MNQ \cong \angle QPM$, and $\angle NMQ \cong \angle PQM$. Prove $\triangle NMQ \cong \triangle PQM$.

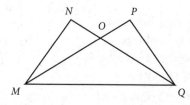

4. Diagonal $\overline{AC}$ is drawn in rectangle $ABCD$. $\angle BAC \cong \angle DCA$. Prove $\triangle ABC \cong \triangle CDA$.

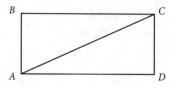

5. In question 4, the quadrilateral $ABCD$ was a rectangle. Suppose we create a similar exercise by drawing diagonal $AC$ in rhombus $ABCD$ and giving $\angle BAC \cong \angle DCA$. Would it still be possible to prove $\triangle ABC \cong \triangle CDA$? If yes, would the argument be different in any way? If no, what additional information would be needed?

6. If $\angle NMO \cong \angle PQO$ and $\overline{NM} \cong \overline{PQ}$ in the figure below, prove $\triangle NMO \cong \angle PQO$.

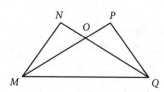

7. $\triangle ABC$ and $\triangle RST$ are isosceles triangles. $\angle B$ is the vertex angle of $\triangle ABC$, the angle included between the congruent sides of $\triangle ABC$, and $\angle S$ is the vertex angle of $\triangle RST$. $\angle B \cong \angle S$ and $\overline{AC} \cong \overline{RT}$. Do you have sufficient evidence to prove $\triangle ABC \cong \triangle RST$? Explain your thinking.

# CPCTC

**CPCTC =** Corresponding Parts of Congruent Triangles are Congruent

When you prove triangles congruent using one of the shortcuts described in lessons 2 and 3, you demonstrate that three pairs of corresponding parts are congruent. You don't have to show that for all six pairs. But the definition of congruence says that all corresponding parts are congruent, so once you've established that the triangles are congruent, all the corresponding parts, even the ones you haven't talked about yet, are congruent.

## EXERCISE 7.4

**All these exercises focus on taking an extra step after proving triangles congruent and showing that a pair of corresponding parts are also congruent. Write a clear proof for each exercise.**

1. $\triangle ABC$ is an isosceles triangle with $\overline{AC} \cong \overline{BC}$. Segment $\overline{CD}$ bisects $\angle ACB$. Prove $\angle CAB \cong \angle CBA$.

2. In the figure below, $\overline{PR}$ and $\overline{SQ}$ bisect one another. Prove $\overline{PS} \cong \overline{QR}$.

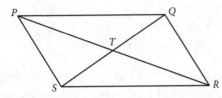

3. In the figure below, $\angle BAE \cong \angle DEA$ and $\angle BEA \cong \angle DAE$. Prove $\overline{AB} \cong \overline{ED}$.

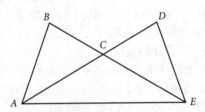

4. In the figure below, $\overline{BD}$ is the bisector of $\angle ABC$ and $\overline{BA} \cong \overline{BC}$. Find the lengths $AB$, $AD$, and $AC$.

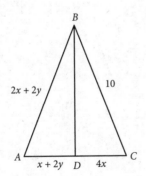

5. In $\triangle PQR$ below, $\overline{RS} \perp \overline{PQ}$ and $\overline{PT} \perp \overline{QR}$. $\angle SRP \cong \angle TPR$. Prove $\overline{SP} \cong \overline{TR}$.

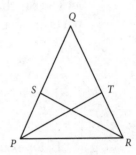

6. In the figure from question 5, the length $SR = 3x + 7$, $PT = 5x - 9$, and $PR = 7x - 1$. Find the length of PR.

7. If you wished to prove the $\overline{BE} \cong \overline{CE}$ in the figure below, which triangles would you prove congruent so that you could use CPCTC? If you knew that $\triangle ABC \cong \triangle DCB$, what would the rest of your argument be to prove $\overline{BE} \cong \overline{CE}$?

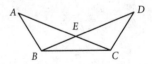

# Multistep Proofs

CPCTC allows you to use the fact that those additional corresponding parts are congruent to prove other statements. Sometimes you want to prove a pair of triangles congruent so that you can show that corresponding parts of those triangles are congruent. Other times you don't have enough information for the proof you really want to do, but proving another pair of triangles first will let you say that corresponding parts of those triangles are congruent, and those parts are shared with your triangles of interest.

EXAMPLE

> If, for example, you wanted to prove $\triangle PQR \cong \triangle TSR$, it might be helpful to first prove $\triangle RXP \cong \triangle RXT$, because it might give you a way to show $RP \cong RT$. Suppose $RX$ is the perpendicular bisector of $PT$, $QR \cong RS$, and $\angle QRP \cong \angle SRT$. The task is to prove $\triangle PQR \cong \triangle TSR$, and we have one pair of congruent sides ($QR \cong RS$) and one pair of congruent angles ($\angle QRP \cong \angle SRT$). We need either another pair of sides or another pair of angles. The other given information shifts attention to $\triangle RXP$ and $\triangle RXT$. If $RX$ is perpendicular to $PT$, $\angle RXP$ and $\angle RXT$ are right angles, and all right angles are congruent, so $\angle RXP \cong \angle RXT$. If $RX$ bisects $PT$, $PX \cong XT$. Add the fact that $RX$ is a side of both $\triangle RXP$ and $\triangle RXT$, and is congruent to itself, so $\triangle RXP \cong \triangle RXT$ by SAS. Then you can say that $RP \cong RT$ and you have the other pair of sides you need. $\triangle PQR \cong \triangle TSR$ by SAS.

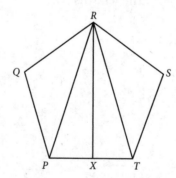

## EXERCISES 7-5

**Take time to plan your proof before you begin writing. Know which triangles you need to prove congruent in order to get the information you need for the task set out in the exercise.**

**1.** If $\overline{AC} \cong \overline{BD}$, $\overline{AE} \cong \overline{DE}$, and $\angle EAC \cong \angle EDB$, prove $\triangle AEB \cong \triangle DEC$.

**5.** If $\angle ABC \cong \angle DCB$, $\angle BCA \cong \angle CBD$, prove $\triangle AEB \cong \triangle DEC$.

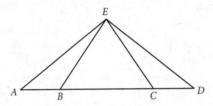

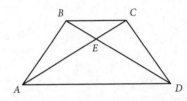

**2.** If $\overline{BC} \cong \overline{CD}$, $\overline{AG} \cong \overline{FE}$, $\angle AGC \cong \angle EFC$, and $\overline{CG} \cong \overline{CF}$, prove $\overline{AB} \cong \overline{ED}$.

**6.** If $\overline{BF} \perp \overline{FE}$, $\overline{CE} \perp \overline{FE}$, $\overline{AB} \cong \overline{CD}$, $\angle FAB \cong \angle EDC$, and $\overline{AF} \cong \overline{DE}$, prove $\triangle FBE \cong \triangle ECF$.

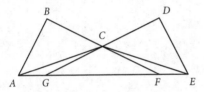

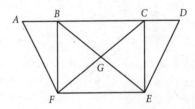

**3.** If $\overline{TS} \cong \overline{RS}$, $\overline{PS} \cong \overline{QS}$, and $\angle RPQ \cong \angle TQP$, prove $\triangle PTQ \cong \triangle QRP$.

**7.** If $\angle L \cong \angle R$, $\angle P \cong \angle N$, and $\overline{LO} \cong \overline{RO}$, prove $\overline{MN} \cong \overline{PQ}$.

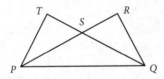

**4.** In the figure below, $ABCD$ is a square. $M$ is the midpoint of $\overline{AD}$ and $N$ is the midpoint of $\overline{BC}$. If $\angle MBD \cong \angle NDB$, prove $\triangle MBD \cong \triangle NDB$.

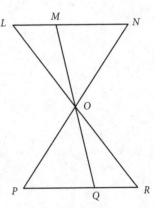

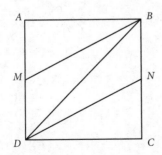

## CHAPTER 8

# Parallelograms and Other Polygons

The study of triangles and their relationships provides a tool for investigating other figures: quadrilaterals of various types and polygons with more than four sides. All that you've learned about congruence will be used again and again.

## Parallelograms

A **parallelogram** is a quadrilateral with two pairs of opposite sides parallel. If you focus on one pair of parallel lines and view the other lines as transversals, you can show that consecutive interior angles are supplementary:

▶ Consecutive angles of a parallelogram are supplementary.

Drawing a diagonal in a parallelogram divides it into two triangles that can easily be shown to be congruent. By CPCTC, it can be proven that every parallelogram has these properties:

▶ Opposite sides of a parallelogram are congruent.

▶ Opposite angles of a parallelogram are congruent.

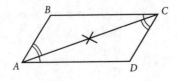

If both diagonals are drawn, proving congruent triangles and showing corresponding parts are congruent will let you show:

▶ The diagonals of a parallelogram bisect each other.

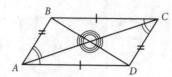

To prove that a quadrilateral is a parallelogram, you need to prove one of the following:

▶ Both pairs of opposite sides are parallel.

▶ Both pairs of opposite sides are congruent.

▶ One pair of opposite sides is both parallel and congruent.

▶ Both pairs of opposite angles are congruent.

▶ Diagonals bisect one another.

The area of a parallelogram is equal to the length of the base times the height, or perpendicular distance between the parallel sides. ($A = bh$)

**EXAMPLE**

▶ $\square ABCD$ has $AB = CD = 12$ cm and the perpendicular distance between $AB$ and $CD$ is 5 cm, the area of $\square ABCD$ is $A = bh = 12 \cdot 5 = 60$ square centimeters.

## EXERCISE 8.1

**Apply properties of parallelograms to solve these questions.**

1. $ABCD$ is a parallelogram with diagonal $AC$. $M$ is the midpoint of $AB$ and $N$ is the midpoint of $CD$. Prove $MN$ is parallel to $AD$.

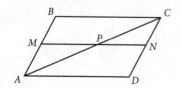

2. The diagonals of parallelogram $ABCD$ intersect at point $E$. $AE = x + 2$, $BE = y - 1$, $CE = 3x - 1$, and $DE = 2y - 3$. Find the length of diagonal $AC$ and diagonal $BD$.

3. $PQRS$ is a parallelogram. $m\angle Q = 11x + 13$ and $m\angle S = 15x - 23$. Find the measure of $\angle P$.

**4.** In the figure below, $\angle BAC \cong \angle ECD$, $\angle CED \cong \angle ECD$, and $ED \cong AB$. Prove $ABCD$ is a parallelogram.

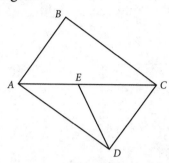

**5.** In the figure below, $\overline{AF} \cong \overline{CG}$, $\angle AFB \cong CGD$, and $\angle BAE \cong \angle DCE$. Prove $ABCD$ is a parallelogram.

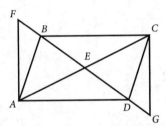

**6.** Find the area of a parallelogram with perimeter of 26 cm if the short side is $x$ cm, the long side is $x + 3$ cm, and the altitude drawn to the long side is $x - 1$ cm.

**7.** In parallelogram $WXYZ$, the area is calculated using the longer side and the altitude drawn to the longer side. If the shorter side is half the longer side, how does the altitude drawn to the shorter side compare to the altitude to the longer side?

**8.** $ABCD$, $EFGH$, and $EICJ$ are parallelograms. Prove $\angle A \cong \angle G$.

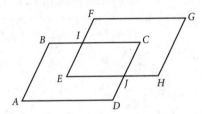

# Rectangles, Rhombuses, and Squares

A **rectangle** is a parallelogram with a right angle. Because every rectangle is a parallelogram, it has all the properties of a parallelogram. Because opposite angles are congruent and consecutive angles are supplementary, all four angles in a rectangle are right angles. It is also possible to show that the diagonals of a rectangle are congruent.

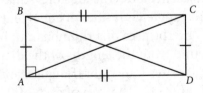

**EXAMPLE**

In rectangle $ABCD$, $AB = CD = 4$ and $AD = BC = 8$. Because all the angles of the rectangle are right angles, the Pythagorean theorem can be applied. The diagonals $BD = AC = \sqrt{4^2 + 8^2} = \sqrt{16 + 64} = \sqrt{80} = 4\sqrt{5}$.

A **rhombus** is a parallelogram with four sides of equal length. It is a parallelogram, so has all the properties of a parallelogram, including opposite angles congruent and consecutive angles supplementary, but may not have all angles congruent. The diagonals of a rhombus intersect at right angles and bisect the interior angles.

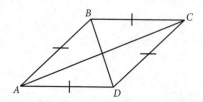

A **square** is a parallelogram that is both a rectangle and a rhombus. A square has four congruent sides and four right angles. It has all the properties of a parallelogram, as well as the properties of rectangles and rhombuses. Because a square is both equilateral and equiangular, it is a regular polygon.

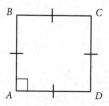

To prove that a quadrilateral is a rectangle or rhombus or square, you must first prove that it is a parallelogram. Once that is established, you need to prove an additional property to classify the parallelogram further.

**Rectangle**     Prove one angle is a right angle, or prove diagonals are congruent.

**Rhombus**     Prove adjacent sides congruent, or prove diagonals are perpendicular or prove the diagonals bisect interior angles.

**Square**     Prove that the parallelogram is both a rectangle and a rhombus.

The area of a rectangle, like any parallelogram, is $A = bh$ but in a rectangle, the base and height are adjacent sides. The area of a square is also $A = bh$, but since the square is a rectangle with sides of equal length, that formula becomes $A = bh = s \cdot s = s^2$. The area of a rhombus can be found as $A = bh$ but can also be found from the lengths of the diagonals. Drawing both diagonals, $d_1$ and $d_2$, creates four right triangles, and each right triangle has legs that measure $\frac{1}{2}d_1$ and $\frac{1}{2}d_2$. The area of each of those right triangles is $A = \frac{1}{2}bh = \frac{1}{2}\left(\frac{1}{2}d_1\right)\left(\frac{1}{2}d_2\right) = \frac{1}{8}d_1d_2$. There are four right triangles making up the rhombus, so the area of the rhombus is $4\left(\frac{1}{8}d_1d_2\right) = \frac{1}{2}d_1d_2$.

In a rhombus with diagonals that measure 25 cm and 30 cm, the area is $\frac{1}{2}d_1d_2 = \frac{1}{2}(25)(30) = 375$ square centimeters.

## EXERCISE 8.2

**Use properties of quadrilaterals and area formulas to answer the questions below. Be sure to include appropriate units.**

1. In quadrilateral $ABCD$, $ABCF$ and $DCBE$ are parallelograms. $AB \cong DC$, and $BC \cong FE$. Prove that $FBCE$ is a rectangle.

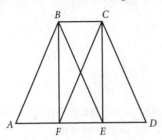

2. $HADE$ and $GHCD$ are parallelograms. $AB \cong BC$ and $\angle A \cong \angle C$. Prove $HBDF$ is a rhombus.

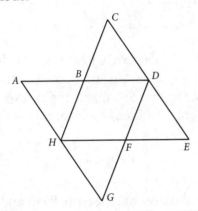

3. $ABCD$ is a parallelogram, $\overline{BE} \cong \overline{ED}$ and $AB \cong BD$. F is the midpoint of $BC$ and E is the midpoint of $AD$. Prove $EBFD$ is a square.

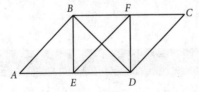

4. The diagonals of a rhombus measure 42 cm and 56 cm. Find the area of the rhombus.

5. A rhombus with a side of 13 cm has a diagonal of 10 cm. A rectangle with the same altitude as the rhombus has an area of 216 cm². Find the base of the rectangle to the nearest tenth of a centimeter.

6. If a square and a rhombus have the same perimeter, can they have the same area? Explain your reasoning.

7. $ABCE$ and $DCBF$ are both rhombuses and share side $BC$. If $\angle A \cong \angle D$, explain how you can conclude that $\triangle BGC$ is isosceles but not equilateral.

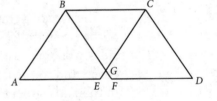

---

# Trapezoids

A **trapezoid** is a quadrilateral with one pair of parallel sides. The parallel sides, usually referred to as the **bases**, are of different lengths. The nonparallel sides, sometimes called *legs*, may be different lengths or the same length. A trapezoid in which the nonparallel sides are congruent is an **isosceles trapezoid**.

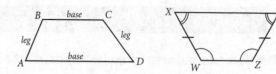

In a trapezoid, the pair of angles at each end of one of the parallel sides is a pair of base angles. In an isosceles trapezoid, a pair of base angles has the same measure. This is not true if the trapezoid is not isosceles. In any trapezoid, the consecutive angles between the parallel sides, along one leg, are **supplementary**.

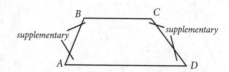

The line segment that connects the midpoints of the nonparallel sides of a trapezoid is the **midsegment**. The midsegment of a trapezoid is parallel to the bases and equal in length to the average of the two bases.

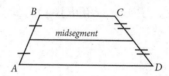

The area of a trapezoid is $A = \frac{1}{2}h(b_1 + b_2)$, where $b_1$ and $b_2$ are the lengths of the two parallel sides and $h$ is the perpendicular distance between the parallel sides. You may notice that $\frac{1}{2}(b_1 + b_2)$ within that formula is the average of the bases, and so the length of the midsegment. The area of a trapezoid is the length of the midsegment times the height.

**EXAMPLE**

If a trapezoid has a height of 12 cm and bases that measure 8 cm and 14 cm, the length of the midsegment is $\dfrac{8 + 14}{2} = \dfrac{22}{2} = 11$ cm and the area of the trapezoid is $11 \cdot 12 = 132$ square centimeters.

## EXERCISE 8.3

**Questions in this group of exercises focus on trapezoids. Use the information in the section to solve each exercise. Include appropriate units.**

1. In $\triangle ABC$, midsegment $\overline{MN}$ connects $M$, the midpoint of $\overline{AB}$, with $N$, the midpoint of $\overline{BC}$. Prove that quadrilateral $AMNC$ is a trapezoid. If the area of $\triangle ABC$ is 64 cm², what is the area of the trapezoid $AMNC$? Explain your reasoning.

2. $ABCD$ is a parallelogram, and $\angle DEC \cong \angle DCE$. Prove quadrilateral $ABED$ is an isosceles trapezoid.

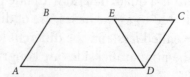

**3.** *ABCD* is an isosceles trapezoid, with $\overline{BC} \parallel \overline{AD}$ and $\overline{AB} \cong \overline{CD}$. Diagonals $\overline{AC}$ and $\overline{BD}$ are drawn. Prove $\overline{AC} \cong \overline{BD}$

**4.** In trapezoid *WXYZ*, $\overline{XY} \parallel \overline{WZ}$. $m\angle W = 2x + 17$, $m\angle X = 5x + 2$, and $m\angle Y = 4x + 3$. Is *WXYZ* an isosceles trapezoid? Explain your reasoning.

**5.** In trapezoid *ABCD*, the midsegment is drawn connecting *M*, the midpoint of $\overline{AB}$, to *N*, the midpoint of $\overline{CD}$. If *MBCN* is rotated 180°clockwise about *N*, what type of quadrilateral is formed? Explain how you would find the area of this quadrilateral and show that it leads to the formula for the area of a trapezoid.

**6.** The key measurements in a trapezoid—the two bases and the height—are known to be 10 cm, 14 cm, and 21 cm, but it is not clear which measurement is the height and which are the bases. Find the area of the trapezoid if (a) h = 10 cm; (b) h = 14; (c) h = 21 cm.

**7.** A trapezoid has an area of 175 cm². If the height is 7 cm and one base measures 11 cm, find the length of the other base.

**8.** Find the height of a trapezoid with bases of 8 in and 12 in, if the area of the trapezoid is 100 in².

# Regular Polygons

A **polygon** is any closed figure made of line segments that intersect only at their endpoints. A polygon is **convex** if any diagonal falls inside the polygon. A polygon that is not convex is **concave**. Concave polygons "cave in," or have dents in them.

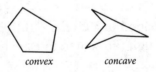

*convex*      *concave*

Polygons are named by number of sides:

3 → triangle
4 → quadrilateral
5 → pentagon
6 → hexagon
7 → heptagon
8 → octagon
9 → nonagon
10 → decagon
12 → dodecagon

A convex polygon is regular if it is both equilateral and equiangular. An equilateral triangle is a regular three-sided polygon because it has three congruent sides and three angles of equal size. Remember a rectangle is equiangular but not equilateral, and a rhombus is equilateral but not equiangular. Only a square is a regular quadrilateral.

Using the construction skills you have learned, it is possible to construct several different regular polygons. These constructions inscribe the regular polygon in a circle, creating the polygon with each of its vertices on a circle.

▶ **Construct a regular hexagon inscribed in a circle.** Draw a radius, connecting the center to any point on the circle. Set the compass opening to the length of the radius of the circle. Place the point of the circle at the point at which the radius meets the circle. Scribe an arc that intersects the circle. Move the point of the compass to the point where the arc crosses the circle, and scribe another arc. Repeat until you return to your starting point. This divides the circle into six congruent arcs. Draw line segments connecting these points in sequence around the circle to create a regular hexagon.

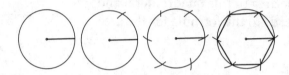

▶ **Construct an equilateral triangle inscribed in a circle.** Begin as for a regular hexagon, dividing the circle into six congruent arcs. Draw line segments connecting every other point around the circle to create an equilateral triangle.

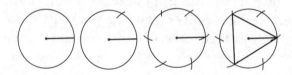

▶ **Construct a square inscribed in a circle.** If the center of the circle is marked, draw a line through the center, intersecting the circle in two points, to create a diameter.

If the center is not marked, construct a right angle with its vertex on the circle by first drawing a line that intersects the circle in two places and then constructing a perpendicular to the line at one of the points where it intersects the circle. Connect the points where the sides of this right angle intersect the circle. That produces a diameter.

Whichever method you used to get a diameter, construct the perpendicular bisector of your diameter to create a second diameter, perpendicular to the first. The points at which these diameters intersect the circle divide it into four congruent arcs. Connect those points with line segments to create a square.

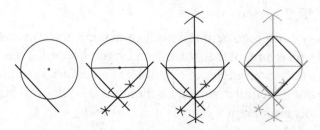

## EXERCISE 8.4

**These questions focus on properties of polygons and constructions of regular polygons. Use the properties of polygons to answer each question.**

**1.** *ABCDEF* is a regular hexagon. Prove △*BDF* is equilateral.

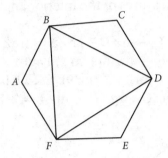

**2.** Identify each figure below. Is it a polygon? Convex or concave? Name it by the number of sides.

(a)      (b)    (c)      (d)      (e)

**3.** The figure below is a 12-sided polygon. All its sides are the same length. Is it a regular dodecagon? Explain your reasoning.

**4.** The construction of a regular hexagon inscribed in a circle divides the circle into six arcs of equal measure. If two arcs of a circle are congruent, the central angles that intercept those arcs are congruent. Explain why, in the diagram below, it is possible to prove $AB \cong CD$.

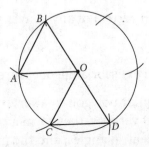

**5.** The construction of a square inscribed in a circle is based on dividing the circle into four congruent arcs. In the figure below, the circle is divided into four arcs, not all congruent. Is it possible to prove: $AB \cong CD$? $AD \cong BC$? $AB \cong AD$? What type of parallelogram is inscribed?

**6.** Use a compass and a straightedge to construct an equilateral triangle within a regular hexagon in circle *O*.

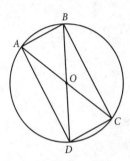

**7.** Use a compass and a straightedge to construct a square and an octagon in circle *O*.

# Angles of Regular Polygons

Like triangles and quadrilaterals, polygons with more than four sides have as many vertices and interior angles as sides. Each vertex is connected to two other adjacent vertices by sides. By drawing diagonals from a vertex to each of the non-adjacent vertices, you can divide a convex polygon into $n - 2$ nonoverlapping triangles. The sum of the measures of the interior angles of those triangles is equal to the sum of the measures of the interior angles of the polygon.

In the figure below, $m\angle A + m\angle B + m\angle C + m\angle D + m\angle E = (m\angle 1 + m\angle 5) + (m\angle 2) + (m\angle 3 + m\angle 4 + m\angle 8) + (m\angle 9) + (m\angle 6 + m\angle 7) = (m\angle 1 + m\angle 2 + m\angle 3) + (m\angle 4 + m\angle 5 + m\angle 6) + (m\angle 7 + m\angle 8 + m\angle 9)$. Each triangle has angles that total 180°, so $m\angle A + m\angle B + m\angle C + m\angle D + m\angle E = 3(180°)$.

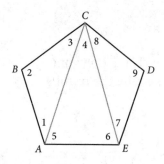

▶ A convex polygon of $n$ sides has $n$ interior angles whose measures add to $180(n - 2)°$.

▶ If the polygon is regular, the measure of each interior angle is $\dfrac{180(n - 2)°}{n}$.

If one side of the polygon is extended through a vertex, an exterior angle is formed. Each exterior angle is supplementary to the interior angle at that vertex. If one exterior angle is created at each vertex, the sum of their measures is equal to:

$(180° - m\angle A) + (180° - m\angle B) + (180° - m\angle C) + (180° - m\angle D)$
$\quad + (180° - m\angle E)$

$= (180° + 180° + 180° + 180° + 180°) - (m\angle A + m\angle B + m\angle C + m\angle D + m\angle E)$

$= 5(180°) - 3(180°)$

$= 2(180°) = 360°$

▶ The sum of the exterior angles (one at each vertex) of any convex polygon is 360°.

▶ If the polygon is regular with $n$ sides, each exterior angle measures $\dfrac{360°}{n}$.

A segment drawn from the center of a regular polygon to a vertex is a radius. When radii are drawn to two adjacent vertices, they form a central angle. A regular polygon with $n$ sides has $n$ central angles, and the measure of each is $\dfrac{360°}{n}$. Drawing all $n$ radii divides the regular polygon into $n$ isosceles triangles. Each triangle has a central angle as its vertex angle, and each of its base angles are half of an interior angle of the regular polygon.

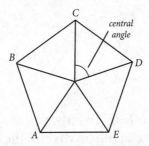

## EXERCISE 8.5

**Calculate the measures of the angles requested using the rules above. If necessary, round to the nearest tenth of a degree.**

1. Find the sum of interior angles of nonagon.

2. Find the measure of one interior angle of a regular octagon.

3. Find the measure of one exterior angle of a regular decagon.

4. For which regular polygons is the sum of the exterior angles greater than or equal to the sum of the interior angles?

5. The sum of the interior angles of a convex polygon is 1,980°. How many sides does the polygon have?

6. If one interior angle of a regular polygon measures 162°, how many sides does the regular polygon have?

7. If one exterior angle of a regular polygon measures 72°, how many sides does the regular polygon have?

# Areas of Regular Polygons

The area of a convex polygon with more than four sides can sometimes be found by breaking it into nonoverlapping triangles, finding the area of each, and summing the results. With regular polygons, this can be accomplished by drawing segments, called **radii**, from the center of the regular polygon to each vertex. In a regular polygon with $n$ sides, this creates $n$ congruent isosceles triangles. If you can find the area of one of these, multiplying by the number of sides will give you the area of the regular polygon.

The formula for the area of a regular polygon is a simplified rule for that process. The area of one triangle is $A = \frac{1}{2}bh$, with $b$ equal to one side of the polygon and $h$ equal to the height or altitude of the small isosceles triangle. That height is called the **apothem** of the regular polygon. The area of the regular polygon with $n$ sides is $A = n\left(\frac{1}{2}bh\right) = \frac{1}{2}(nb)h$. Because $nb$ is the perimeter of the regular polygon and $h$ is the apothem, that formula can be represented as

$A = \frac{1}{2}aP$, where $a$ is the length of the apothem and $P$ is the perimeter of the regular polygon.

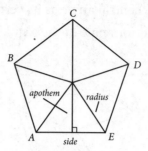

If you are not given the apothem directly, you may be able to find it from other information. Because the apothem divides the small isosceles triangle into two right triangles, if you know the side and the radius, you can find the apothem using the Pythagorean theorem.

If you don't have enough information to do that, calculate the measures of the angles in the small isosceles triangle and see if you have a special right triangle and can find the length of the apothem from that. If all that fails, you'll need to use trigonometry.

Trigonometry comes up later in this book, but if you have a calculator with a key that says *tan*, you can find the apothem by multiplying half the length of a side times tan(measure of base angle of the little isosceles triangle).

**EXAMPLE**

> In a regular pentagon with a side of 6 cm, the little isosceles triangle has a base angle of 54°, so the apothem is 3tan(54) ≈ 4.13 cm.

## EXERCISE 8.6

**Perform the requested calculations according to the rules above. Be sure to include appropriate units.**

1. The radii of a regular pentagon are drawn, dividing the pentagon into five isosceles triangles. Find the measures of the three angles of one isosceles triangle.

2. A regular octagon has an apothem that measures 12 cm and a radius of 13 cm. Find the perimeter of the octagon.

3. A regular hexagon has a radius of 10 cm. Find the length of a side of the hexagon and the length of the apothem.

4. If all radii are drawn in a regular hexagon with a perimeter of 48 cm, find the area of one of the triangles created.

5. If all radii are drawn in a regular pentagon with a perimeter of 40 cm and apothem of 5.5 cm, find the area of one of the triangles created.

6. Find the area of a regular octagon with a side of 10 cm and a radius of 13 cm.

7. Find the area of a regular decagon with a side of 15 cm and an apothem of 23 cm.

8. Which has the larger area: regular hexagon with a side of 8 cm or a regular octagon with a side of 6 cm and an apothem of 7.24 cm?

# Congruence, Parallelograms, and Polygons

The questions in this review section bring together the ideas from Chapters 7 and 8. They will give you a chance to see if you have mastered the idea of congruence and can solve problems about polygons. Answer all the questions, and try to express your thinking as clearly as you can.

1. In the figure below, $\overline{AB} \parallel \overline{DC}$. Is it possible to prove $\triangle AOB \cong \triangle DOC$? Explain.

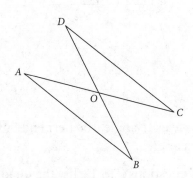

**2.** In the figure below, $\overline{XZ}$ bisects $\angle WZY$ and $\angle WXY$. Is it possible to prove $\triangle WZX \cong \triangle YZX$?  Explain.

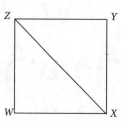

**3.** In the figure below, $O$ is the midpoint of $\overline{AC}$, and $\overline{AB} \parallel \overline{DC}$. Prove $\triangle AOB \cong \triangle COD$.

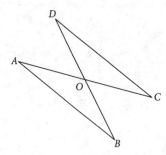

**4.** In the figure below, $\overline{TR}$ bisects $\angle PRQ$ and $\angle PTR \cong \angle QTR$. Prove $\overline{PT} \cong \overline{QT}$.

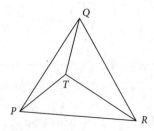

**5.** In the figure below, $\overline{AB} \cong \overline{CD}$ and $\angle ABC \cong \angle DCB$. Prove $\overline{AC} \cong \overline{DB}$. Continue on to prove $\overline{AE} \cong \overline{ED}$.

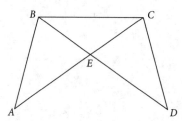

**6.** Prove that in any isosceles triangle, the perpendicular bisector of the base also bisects the vertex angle.

For each property in questions 7 to 11, list the quadrilateral(s) that have that property: parallelogram, rectangle, rhombus, square, trapezoid, isosceles, trapezoid.

**7.** Diagonals are congruent.

**8.** Diagonals are perpendicular.

9. Diagonals bisect one another.

10. Diagonals bisect interior angles.

11. Midsegment is parallel to bases.

12. Given that $M$ is the midpoint of $\overline{AC}$ and the midpoint of $\overline{BD}$, prove that $AD \cong BC$.

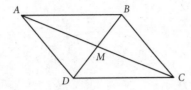

13. Given that *HOME* is a parallelogram, find all possible values of $x$ and $y$.

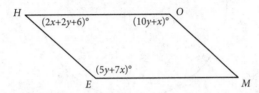

14. If *ABCD* and *AEFG* are parallelograms, prove that $\angle D$ is supplementary to $\angle F$.

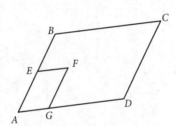

15. On the space below, graph the quadrilateral with vertices $P(-6, 0)$, $Q(0, 3)$, $R(2, -1)$, $S(-4, -4)$ and prove that *PQRS* is a rectangle. Explain your reasoning.

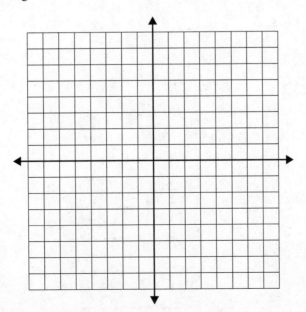

16. Each interior angle of a regular polygon measures 144°. How many sides does the polygon have?

17. Find the number of degrees in one interior angle of a regular heptagon.

18. Find the sum of the interior angles in a 15-sided polygon.

19. Use compass and straightedge to inscribe a regular hexagon in the circle below.

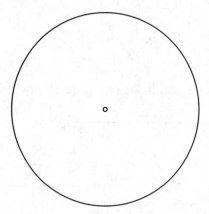

20. Find the area of the hexagon in question 19 if the radius of the circle is 12 cm.

## CHAPTER 9

# Similarity

As mentioned in Chapter 7, the idea of *congruence*, the correspondence between two polygons of exactly the same size and shape, is one of the key relationships in geometry. The other is **similarity**, which finds correspondence between figures that are the same shape but different sizes. When a rigid transformation is applied to a polygon, it produces an image that is congruent to the pre-image. An image similar to the pre-image is created by applying a transformation called a dilation.

## Dilations

A **dilation** is a transformation that produces an image which is the same shape as the pre-image, but a different size. If the image is larger than the pre-image, the dilation is an enlargement; if the image is smaller, the dilation is a reduction.

One way to visualize a dilation is to imagine a projector directing beams of light toward an image. The greater the distance between the projector and the surface on which the image falls, the larger the image will appear. The dilation is described by its **center**, a fixed point akin to the location of the projector, and by the **scale factor**, the ratio of image size to pre-image size. Enlargements have scale factors greater than 1, while reductions have scale factors less than 1. A scale factor of 3 indicates that the image is three times the size of the pre-image. A scale factor of ½ means the image is half the size of the pre-image.

Think of the center as the projector's light source, sending beams of light toward the pre-image. It can be a point on the polygon or elsewhere in the plane. Focus your attention on the vertices of the polygon, which determine its shape. The beams of light that strike those vertices diverge as they continue on. A scale factor of 2 means that $PA' = 2 \cdot PA$. A scale factor less than 1 means the image is closer to the center, and therefore smaller.

EXAMPLE

If the center of the dilation is $P$ in the figure below, the rays through $A$, $B$, and $C$ are extended and $A'$, $B'$, and $C'$ are measured out on the corresponding rays.

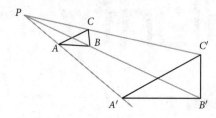

Like rigid transformations, dilations are often easier to draw if you work on a coordinate plane. In the figure below, $\triangle PQR$ is subjected to a dilation with a scale factor of $\frac{1}{3}$, centered at the origin. The distance from the origin to a point on the image, like $R'$, is $\frac{1}{3}$ of the distance from the origin to the corresponding point $R$ on the pre-image, but notice that in the image, the coordinates of each vertex of the image are $\frac{1}{3}$ of the coordinates in the pre-image.

To perform a dilation centered at the origin, multiply the coordinates of the pre-image by the scale factor.

EXAMPLE

In the figure below $\triangle PQR$ is subject to a dilation centered at the origin with a scale factor of $\frac{1}{3}$. $P(0, 3)$ is sent to $P'(0, 1)$, $Q(3,9)$ to $Q'(1, 3)$, and $R(6, 6)$ to $R'(2, 2)$.

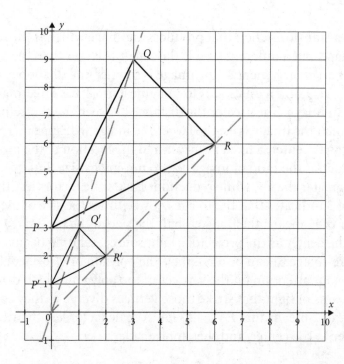

If the center of the dilation is not the origin, you can translate the center to the origin, apply the same translation to the pre-image, multiply by the scale factor, draw the image, and, finally, translate everything back.

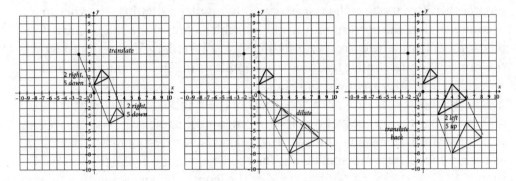

## EXERCISE 9.1

**Use properties of dilations to answer each of the following exercises. Draw the pre-image and image on a coordinate plane if helpful.**

1. A rectangle with sides of 6 cm and 17 cm is subject to a dilation by a factor of 0.7. What is the perimeter of the image?

2. $\triangle RST$ has vertices $R(3,-4)$, $S(5, -1)$, and $T(2, 4)$. If a dilation by a factor of 2 centered at the origin is applied, draw the image and give its coordinates.

3. A rectangle with vertices $A(-2,2)$, $B(4, 2)$, $C(4, -4)$, and $D(-2, -4)$ is subjected to a dilation by a factor of 1.5, centered at the origin. Draw the image and give its coordinates.

4. If a dilation centered at the origin is applied to the image rectangle $A'B'C'D'$ in exercise 3, what scale change is required to carry it back onto its pre-image $ABCD$?

5. Apply a dilation with scale factor 2.5, centered at $R(3,-4)$ to $\triangle RST$ with vertices $R(3,-4)$, $S(5, -1)$, and $T(2, 4)$.

6. Apply a dilation centered at $P(2, -3)$ with a scale factor of $\frac{1}{3}$ to the rectangle $ABCD$ with coordinates $A(-1,3)$, $B(8, 3)$, $C(8, -9)$, and $D(-1, -9)$.

7. Scale changes greater than 1 enlarge the pre-image, and scale changes less than 1 reduce it. A scale change of zero would reduce the entire image to one point at the origin, not a very useful change. If you were asked to apply a dilation by a negative factor, how could you make sense of that request? What would the image look like?

# Similarity

Two polygons are similar if they are the same shape but different sizes, or put another way, if one is the image of another under a dilation. In order to accomplish this, the corresponding angles, which control the shape, must be congruent, and the corresponding sides must be enlarged or reduced by the same factor. In similar figures, corresponding angles are congruent and corresponding sides are in proportion.

The symbol for "is similar to" is ~, part of the symbol for "is congruent to," ≅. Similar figures are the same shape (~) but not the same size (=). The way in which the similarity statement is written makes the correspondence clear, just as congruence statements spelled out the correspondence. If $\triangle ABC \sim \triangle YXZ$, then $\angle A \cong \angle Y$, $\angle B \cong \angle X$, and $\angle C \cong \angle Z$. The lengths of corresponding sides create the proportion $\dfrac{AB}{YX} = \dfrac{BC}{XZ} = \dfrac{AC}{YZ}$.

## EXERCISE 9.2

These exercise focus on the definition of similarity and the concept of proportional sides. Use that information to complete each question.

1. If $\triangle XYZ \sim \triangle RTS$, list the pairs of corresponding parts.

2. How is the similarity statement $\triangle XYZ \sim \triangle RTS$ different from the similarity statement $\triangle XYZ \sim \triangle RST$?

3. In $\triangle ABC$, $m\angle A = 43°$ and $m\angle B = 112°$. In $\triangle XYZ$, $m\angle X = 25°$ and $m\angle Y = 43°$. Could the triangles be similar? If so, write the correct similarity statement. If not, explain why.

4. $\triangle RST$ has sides measuring 12 cm, 30 cm, and 22 cm. $\triangle DEF$ has sides of 12 cm, 6 cm, and 18 cm. Could the triangles be similar? If so, write the correct similarity statement. If not, explain why.

5. In one triangle, $AR = 18$ cm, $RM = 25$ cm, and $AM = 40$ cm. In another triangle, $GL = 60$ cm, $EG = 27$ cm, and $EL = 37.5$ cm.

Write the similarity statement and give the scale factor.

6. Write the correct similarity statement for the similar pentagons below. Explain how you determined which parts correspond.

7. A rectangle has a width of 6 inches and a length of 10 inches. The rectangle undergoes a dilation with a scale factor of 2.4. Find the perimeter and area of the enlarged rectangle. How do they compare to the perimeter and area of the original rectangle?

# Proving Triangles Similar

Just as it is not necessary to prove all pairs of corresponding parts congruent in order to claim two triangles are congruent, it is not necessary to prove all corresponding angles congruent and all corresponding sides in proportion in order to show two triangles are similar. The most common method of showing two triangles are similar is "angle-angle."

▶ **AA**   If two angles of one triangle are congruent to the corresponding angles of the other triangle, then the triangles are similar.

It is not necessary to prove that the third pair of angles are congruent, because the Triangle Sum theorem guarantees that the sum of the three angles in any triangle is $180°$, and so it takes little more than arithmetic to prove the Third Angle theorem: If two angles of one triangle are congruent to the corresponding angles of another triangle, then the third angles will also be congruent.

Other less commonly used shortcuts for showing two triangles similar involve showing some or all sides are in proportion.

▶ **SSS**   If all three pairs of corresponding sides in one triangle are in proportion to the corresponding sides of another triangle, the triangles are similar.

▶ **SAS**   If two sides of one triangle are proportional to the corresponding sides of the other triangle and the angles included between those sides are congruent, then the triangles are similar.

After proving a pair of triangles congruent, you often needed to use the fact that corresponding parts of congruent triangles are congruent (CPCTC) to prove that a pair of angles or a pair of sides, not previously mentioned, was congruent. When you prove triangles similar, you may need to make an additional claim as well. It is not usually a claim about angles, but rather one about sides being in proportion. The reason you can claim that the proportion is true is "corresponding sides of similar triangles are in proportion," but that doesn't usually get a handy abbreviation.

## EXERCISE 9.3

**Prove triangles similar using the method appropriate to the given information. Show the steps in your argument, and provide reasons for the claims you make.**

1. Begin with $\triangle ABC$ and $\triangle XYZ$, and the given information that $\angle A \cong \angle X$ and $\angle B \cong \angle Y$. Prove the Third Angle theorem by showing that $\angle C$ must be congruent to $\angle Z$.

2. In the figure below, $\overline{AB} \parallel \overline{DE}$. Prove $\triangle ABC \sim \triangle DEC$.

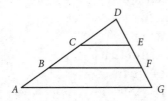

3. In the figure below, $\overline{AB} \parallel \overline{DF}$. Prove $\triangle ABC \sim \triangle DEC$.

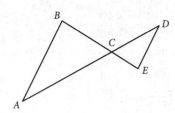

4. In the figure below, $\overline{CE} \parallel \overline{BF} \parallel \overline{AG}$. Prove $\triangle CDE \sim \triangle BDF$.

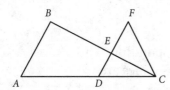

5. In the figure below, $\overline{RT} \perp \overline{PS}$ and $\angle PQT \cong \angle SRT$. Prove $\dfrac{PQ}{SR} = \dfrac{PT}{ST}$.

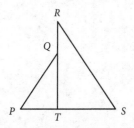

6. In the figure below, $\angle X \cong \angle V$. Prove $\dfrac{WY}{ZY} = \dfrac{WX}{ZV}$

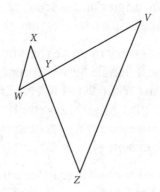

7. Altitude $\overline{AB}$ is drawn in $\triangle ADC$ and altitude $\overline{JK}$ is drawn in $\triangle JML$. If $\triangle ACD \sim \triangle JLM$, prove $\triangle ABC \sim \triangle JKL$.

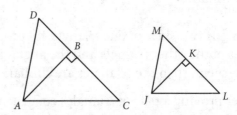

8. $\triangle PQR$ below is isosceles with $\overline{PQ} \cong \overline{QR}$ and $\triangle RST$ is isosceles with $\overline{RS} \cong \overline{ST}$. If $T$ is the midpoint of $\overline{PR}$ and $\overline{PQ} \parallel \overline{SR}$, prove that $RS = \frac{1}{2}PQ$.

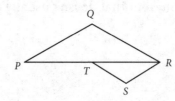

# Using Proportions

Knowing that corresponding sides of similar triangles are proportional can be helpful in finding missing dimensions. If you can write a proportion about the sides of similar triangles that contains enough known information, you can solve for the missing side(s).

There are a few key things about proportions worth reviewing. A proportion is a statement that says two ratios are equal. A proportion is usually written with the ratios as fractions, as in $\frac{8}{3} = \frac{16}{6}$, but could be written on a single line like this: 8:3 = 16:6. The first and last numbers are called the **extremes**, and the two in the middle are called the **means**. The most important property of proportions is the **Means-Extremes property**, which says that the product of the means is equal to the product of the extremes.

**EXAMPLE**

In this example, the extremes are 8 and 6 and the means are 3 and 16. By the Means-Extremes property, $8 \times 6 = 3 \times 16 = 48$. When we write proportions using fractions, this rule usually is called **cross-multiplying**.

$$\frac{8}{3} \bcancel{\times} \frac{16}{6}$$
$$8 \times 6 = 16 \times 3$$

When the proportion involves a variable, cross-multiplying leads to an equation that can be solved.

**EXAMPLE**

$$\frac{7}{x} = \frac{5}{8}$$
$$5x = 7 \times 8$$
$$5x = 56$$
$$x = \frac{56}{5}$$
$$x = 11.2$$

Before we look at other properties of proportions, remember that the denominator of a fraction can never be zero. That won't often come up when you're working with lengths of sides, but be thoughtful when using variables.

> The important thing to remember is to be consistent in writing your ratios. Large to small is fine and so is small to large, but whichever you choose, both ratios must be the same.

▶ If $\frac{a}{b} = \frac{c}{d}$, then $\frac{b}{a} = \frac{d}{c}$. You can invert the ratios. Remember to invert both.

▶ If $\frac{a}{b} = \frac{c}{d}$, then $\frac{d}{b} = \frac{c}{a}$. You can swap the extremes. You can also swap the means.

▶ If $\frac{a}{b} = \frac{c}{d}$, then $\frac{a+b}{b} = \frac{c+d}{d}$ and $\frac{a-b}{b} = \frac{c-d}{d}$. $\frac{a+b}{b} = \frac{c+d}{d}$ can be simplified to $\frac{a}{b} + \frac{b}{b} = \frac{c}{d} + \frac{d}{d}$ or $\frac{a}{b} + 1 = \frac{c}{d} + 1$.

These are just telling you what things look like when you add (or subtract) an equal amount to both sides of the equation.

## EXERCISE 9.4

**Use the concept of proportional sides of similar triangles and the properties of proportions to complete these exercises.**

1. Solve for $x$: $\dfrac{x}{7} = \dfrac{12}{21}$.

2. Solve for $x$: $\dfrac{11}{18} = \dfrac{5}{x+3}$.

3. Show that each of the variations on the proportion $\frac{a}{b} = \frac{c}{d}$, shown in the properties above, ($\frac{b}{a} = \frac{d}{c}, \frac{a+b}{b} = \frac{c+d}{d}, \frac{a-b}{b} = \frac{c-d}{d}$) results in an equivalent equation after the Means-Extremes property is applied.

4. There is a similarity correspondence in $\triangle ABC$ and $\triangle XYZ$, so that $\dfrac{AB}{XY} = \dfrac{BC}{YZ}$. If $AB = 14$, $XY = 49$, and $YZ = 57$, find $BC$.

5. $\triangle A'B'C'$ is an enlargement of $\triangle ABC$ such that each pair of corresponding sides are in ratio of $\frac{9}{4}$. If $AB = 9$ cm, how long is $A'B'$?

6. $\triangle X'Y'Z'$ is a reduction of $\triangle XYZ$ with a scale factor of $\frac{13}{20}$. If $X'Y' = 156$ cm and $XZ = 110$ cm, find the lengths of $XY$ and $X'Z'$.

7. Solve $\dfrac{x}{12} = \dfrac{5}{x+7}$ for a positive value of $x$. If this proportion expressed the proportional relationship of sides of similar triangles, what would the scale factor be?

# Dividing the Sides of a Triangle

A **midsegment** of a triangle is a line segment that connects the midpoints of two sides of a triangle. The **Midsegment theorem** says that the segment that connects the midpoints of two sides of a triangle is parallel to the third side and half as long.

▶ **Midsegment theorem**   The segment that connects the midpoints of two sides of a triangle is parallel to the third side and half as long.

**EXAMPLE**

In $\triangle ABC$, $M$ is the midpoint of $\overline{AB}$ and $N$ is the midpoint of $\overline{BC}$. $M$ divides $\overline{AM}$ into $\overline{MB}$ and $\overline{AB}$, each half as long as $\overline{AB}$, and $N$ divides $\overline{BC}$ into $\overline{BN}$ and $\overline{NC}$, each half of $\overline{BC}$. Therefore, $\dfrac{MB}{AB} = \dfrac{BN}{BC} = \dfrac{1}{2}$. Two pairs of corresponding sides are in proportion. $\angle B$ is included between $\overline{MB}$ and $\overline{BN}$ in $\triangle MBN$ but also included between $\overline{AB}$ and $\overline{BC}$ in $\triangle ABC$. Because $\angle B$ is congruent to itself, $\triangle MBN \sim \triangle ABC$ by SAS. Because the triangles are similar, corresponding angles are congruent, and $\angle A \cong \angle BMN$, and that proves that $\overline{MB} \parallel \overline{AC}$. Like the other sides of the two triangles, $\dfrac{MN}{AC} = \dfrac{1}{2}$.

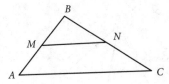

This idea can be generalized to line segments that cut two sides in other ratios. If a line segment connects two sides of a triangle so that the sides are divided in the same ratio $\frac{a}{b}$, then the line segment is parallel to the third side of the triangle and its length is $\frac{a}{a+b}$ of the third side. (See the proof of the Midsegment theorem and replace $\frac{1}{2}$ with $\frac{a}{a+b}$.)

The converse of this generalized version is known as the **Side-Splitter theorem**.

**Side-Splitter theorem**    If a line is drawn parallel to one side of a triangle, it divides the other two sides proportionally.

**EXAMPLE**

If $\overline{DE} \parallel \overline{AC}$, then $\dfrac{BD}{BA} = \dfrac{BE}{BC} = \dfrac{DE}{AC}$ and $\dfrac{BD}{DA} = \dfrac{BE}{EC}$.

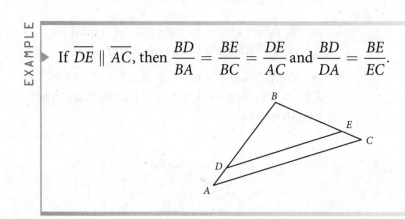

▶ The **Angle Bisector theorem** also talks about dividing a side of a triangle and creating proportional relationships.

▶ **Angle Bisector theorem**   The bisector of an interior angle of a triangle divides the side opposite the angle into segments whose lengths are proportional to the adjacent sides.

▶ If $\overline{BD}$ bisects $\angle ABC$, then $\dfrac{AD}{AB} = \dfrac{DC}{BC}$. The proof of this one requires trigonometry, so we'll save that for later.

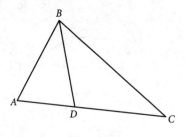

# EXERCISE 9.5

**These questions focus on dividing a single triangle in ways that create similar triangles or proportional relationships. Use the information in this section to complete the exercises.**

1. If $\overline{XY} \parallel \overline{AB}$ in the triangle below, you can prove that $\dfrac{CX}{CA} = \dfrac{CY}{CB}$. Use properties of proportions to show that if $\dfrac{CX}{CA} = \dfrac{CY}{CB}$, then $\dfrac{CX}{XA} = \dfrac{CY}{YB}$.

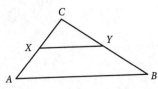

2. In the figure below, $\overline{XY} \parallel \overline{WV} \parallel \overline{AB}$. Prove $\dfrac{XW}{WA} = \dfrac{YV}{VB}$.

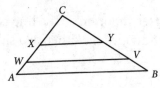

3. In the figure above, is the ratio $\frac{XY}{AB}$ equal to $\frac{CX}{XA}$ or $\frac{CX}{CA}$? Explain your reasoning.

4. In $\triangle WXY$, $\overline{XZ}$ bisects $\angle X$. If $XY = 3$ inches, $XW = 4$ inches, and $WY = 5$ inches, find the length of $\overline{WZ}$.

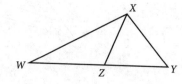

**5.** In $\triangle ABC$, $\overline{BD}$ bisects $\angle B$ and divides $\overline{AC}$ into two segments. $AD = 32$ cm and $DC = 20$ cm. If $AB$ is 15 cm longer than $BC$, find the length of $\overline{BC}$.

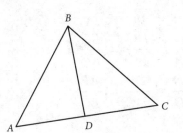

**6.** In $\triangle ABC$ below, $\overline{AX}$, $\overline{BY}$, and $\overline{CZ}$ are angle bisectors and $P$ is the incenter of the triangle. Explain why $\dfrac{AP}{AB} = \dfrac{PX}{BX}$.

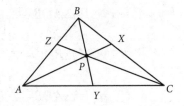

# Altitude to the Hypotenuse

One of the special properties of any right triangle is that each leg is also an altitude. The altitude from the right angle vertex to the hypotenuse is not a side of the original triangle, but it comes with its own special properties.

The altitude to the hypotenuse of a right triangle divides the triangle into two right triangles, each of which is similar to the original triangle. The two smaller right triangles are similar to each other as well:

EXAMPLE

> $\triangle ABC \sim \triangle ADB \sim \triangle BDC$

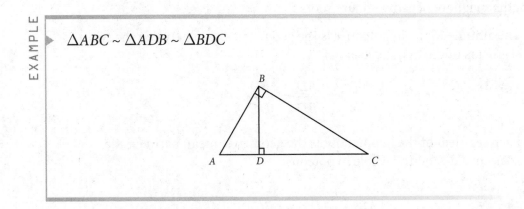

Finding the correct correspondence for the similarity statement can be a challenge however. The right angles must correspond, and those are at vertex $D$ in the small triangles, but vertex $B$ in the original. Then locate the smaller acute angle in each triangle. In $\triangle ADB$, the smallest angle is $\angle B$, but in $\triangle CDB$, and in $\triangle ABC$, it's $\angle C$.

Altitude $\overline{BD}$ to the hypotenuse of $\triangle ABC$ creates two right triangles, $\triangle ADB$ and $\triangle CDB$, which are similar to each other and to the original triangle, under the correspondence $\triangle ABC \sim \triangle ADB \sim \triangle BDC$.

Corresponding sides are in proportion:

▶ If $\triangle ADB \sim \triangle ABC$ $\qquad \dfrac{AD}{AB} = \dfrac{BD}{BC} = \dfrac{AB}{AC}$,

▶ If $\triangle BDC \sim \triangle ABC$ $\qquad \dfrac{BD}{AB} = \dfrac{DC}{BC} = \dfrac{BC}{AC}$,

▶ If $\triangle ADB \sim \triangle BDC$ $\qquad \dfrac{AD}{BD} = \dfrac{DB}{DC} = \dfrac{AB}{BC}$

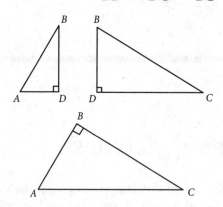

From those proportions come two rules about the lengths of segments, both involving the geometric mean. In a proportion like $\dfrac{a}{b} = \dfrac{c}{d}$, $b$ and $c$ are the means. If both $b$ and $c$ are the same quantity, that quantity is the geometric mean. If $\dfrac{a}{b} = \dfrac{b}{d}$, then $b^2 = ad$ and $b = \sqrt{ad}$. Geometry deals with segment lengths so ignore negative square roots.

▶ The altitude to the hypotenuse is the geometric mean between the two segments of the hypotenuse.

$$\frac{AD}{BD} = \frac{BD}{DC}$$

▶ Each segment of the hypotenuse is the geometric mean between the adjacent side and the whole hypotenuse.

$$\frac{AB}{AD} = \frac{AD}{AC} \qquad \frac{BC}{DC} = \frac{DC}{AC}$$

## EXERCISE 9.6

**These exercises focus on the similar triangles created when an altitude is drawn from the right angle to the hypotenuse of a right triangle. Complete each exercise.**

In right triangle $\triangle ABC$ below, altitude $\overline{BD}$ is drawn to hypotenuse $\overline{AC}$. Use this figure for questions 1 through 5.

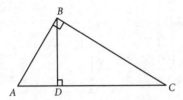

**1.** If $BD = 14$ and $DC = 49$, find $AD$.

**2.** If $AB = 6$ and $AD = 3$, find $DC$.

**3.** If $AD = 8$ and $BD = 12$, find $AC$.

**4.** If $BC = 15$ and $AC = 25$, find $DC$.

**5.** $AD = 8$ and $BD = 16$, find $DC$, $AC$, $AB$, and $BC$.

**6.** Use the diagram below to write the proportions you might use to find the lengths $a$ and $b$. Cross-multiply but do not solve. Use the equations to write an expression for $a^2 + b^2$ in simplest form.

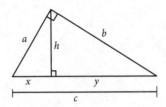

**7.** What theorem does question 6 prove?

## CHAPTER 10

# Right Triangle Trigonometry

*T*rigonometry means "triangle measurement." It uses ideas from similarity to provide a means of finding measurements of sides or angles in right triangles.

## Similar Right Triangles

To be certain that two triangles are similar, all that is necessary is to show that two angles in one triangle are congruent to two angles of the other. Because all right triangles have a right angle, they're halfway to being similar. All that's needed is to find a pair of acute angles, one from each triangle, that are congruent. That means that if you have a right triangle with a 23° angle, it's similar to every other right triangle with a 23° angle. It's as if the universe of right triangles is divided into families of similar right triangles, based on the size of the acute angles. In the figure below, $\triangle ABK \sim \triangle ACJ \sim \triangle ADI \sim \triangle AEH \sim \triangle AFG$. All of these triangles are similar because each has a right angle and each contains $\angle A$.

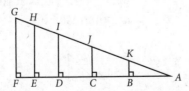

Let's focus on $\triangle ABK$ and $\triangle AFG$, the smallest and largest of the group. Obviously $\angle A \cong \angle A$, and the right angles $\angle ABK$ and $\angle AFG$ are congruent, so by the Third Angle theorem, $\angle AKB \cong \angle AGF$. Corresponding sides

are in proportion, and the proportion for these two triangles would say
$\frac{AB}{AF} = \frac{BK}{FG} = \frac{AK}{AG}$. The last ratio in that proportion is the ratio of the two hypotenuses, and the middle ratio compares the sides opposite $\angle A$. Sides $\overline{AB}$ and $\overline{AF}$, whose lengths are compared in the first ratio, are the sides adjacent to $\angle A$. Of course, $\angle A$ is between two sides, but one is always called the hypotenuse, so the non-hypotenuse side is referred to as the adjacent side. The proportion above says $\frac{\text{adjacent}}{\text{adjacent}} = \frac{\text{opposite}}{\text{opposite}} = \frac{\text{hypotenuse}}{\text{hypotenuse}}$. Whatever scale factor enlarges or reduces the adjacent sides affects the other sides the same way.

If we focus on just two of the ratios at a time, and apply properties of proportions, $\frac{\text{adjacent}}{\text{adjacent}} = \frac{\text{opposite}}{\text{opposite}}$ could become $\frac{\text{opposite}}{\text{adjacent}} = \frac{\text{opposite}}{\text{adjacent}}$, and $\frac{\text{adjacent}}{\text{adjacent}} = \frac{\text{hypotenuse}}{\text{hypotenuse}}$ could become $\frac{\text{adjacent}}{\text{hypotenuse}} = \frac{\text{adjacent}}{\text{hypotenuse}}$. For example, $\frac{AB}{BK} = \frac{AF}{FG}$ says the ratio of the adjacent side to the opposite side in little $\triangle ABK$ is the same as the ratio of the adjacent side to the opposite in big $\triangle AFG$.

There are many different ways of organizing the information about corresponding sides into proportions. The key is to have a plan for what you want to compare, and hold to that for the entire proportion.

## EXERCISE 10.1

**The area of mathematics labeled trigonometry is based in similar triangles. Use properties of similarity to complete these exercises.**

**1.** Write a similarity statement that makes clear the correspondence between the similar right triangles below.

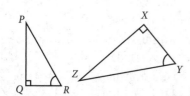

**2.** If $\triangle DOG \sim \triangle CAT$, write the extended proportion that describes the relationships among the sides.

**3.** $\triangle ARM \sim \triangle LEG$. If $AR = 22$ cm, $RM = 17$ cm, and $LE = 33$ cm, find the length of $\overline{EG}$.

**4.** $\triangle PQR \sim \triangle MLK$. If $PQ = 21$ inches, $QR = 25$ inches, and $LK = 5$ inches, find the length of $ML$. What is the scale factor which should be applied to $\triangle PQR$ to produce $\triangle MLK$?

**5.** If the triangles in exercise 4 are right triangles, find the length of each hypotenuse.

**6.** In each of the similar right triangles below, what is the ratio of the leg opposite $\angle A$ to the leg adjacent to $\angle A$?

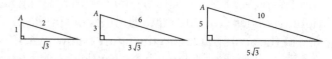

**7.** In each of the similar right triangles below, what is the ratio of the leg opposite $\angle X$ to the hypotenuse?

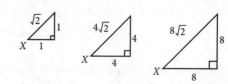

# Basic Ratios

As the last few exercises suggest, a particular ratio of opposite side to hypotenuse, or adjacent side to opposite side, or whatever, turns out to be the same for every triangle in a particular family. If your right triangle has a 60° angle, the ratio of the side to adjacent to the 60° angle to the hypotenuse is $\frac{1}{2}$. It may be that the adjacent side is 3 inches and the hypotenuse is 6 inches, or it may be that the adjacent side is 20 miles and the hypotenuse is 40 miles. It will always reduce to $\frac{1}{2}$. Not every ratio will be such a tidy fraction. If your angle is 53° instead of 60°, the ratio will be an irrational number approximately equal to 0.545, but just as the ratio will be $\frac{1}{2}$ for the entire 60° family, it will be approximately 0.545 for the entire 53° family.

If every family of right triangles has a set value for each of the six possible ratios, then it makes sense to record those values to make them easier to use. First, we name the ratios to make them easier to talk about.

**sine:** $\dfrac{\text{opposite}}{\text{hypotenuse}}$ **cosine:** $\dfrac{\text{adjacent}}{\text{hypotenuse}}$ **tangent:** $\dfrac{\text{opposite}}{\text{adjacent}}$

**cosecant:** $\dfrac{\text{hypotenuse}}{\text{opposite}}$ **secant:** $\dfrac{\text{hypotenuse}}{\text{adjacent}}$ **cotangent:** $\dfrac{\text{adjacent}}{\text{opposite}}$

**EXAMPLE**

> Instead of "for every right triangle in the 60° family, the ratio of the adjacent side to the hypotenuse is $\frac{1}{2}$" we say $\cos(60°) = \frac{1}{2}$. Cosine is the name of that ratio, it's abbreviated as cos, and 60° is the "family."

The values for all these ratios were recorded in tables, and you could look up the values you wanted to use (and you still can.) Now, scientific calculators or graphing calculators can provide the values for you. You'll see keys for *sin*, *cos*, and *tan* on your calculator. If you want the value for the sine, cosine, or tangent for any family of right triangles, those keys will provide that information.

## EXERCISE 10.2

**Use the right triangle below and the definitions of the trig ratios to find the ratios in questions 1 through 6.**

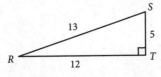

**1.** $\sin(\angle R)$

**2.** $\cos(\angle T)$

**3.** $\sec(\angle T)$

**4.** $\cot(\angle R)$

**5.** $\cot(\angle T)$

**6.** $\tan(\angle R)$

**7.** The trig ratios are defined for the acute angles of a right triangle. What would happen if you tried to apply the definitions to the right angle?

**8.** Use your calculator to find each of the following, rounded to the nearest thousandth.

a) $\sin(30°)$

b) $\cos(45°)$

c) $\tan(82°)$

d) $\cos(11°)$

e) $\sin(75°)$

---

EXAMPLE

The **similar triangles method** says if $\triangle ABC \sim \triangle XYZ$ and you have enough information to set up a proportion, you can solve for an unknown side. If $\dfrac{15}{9} = \dfrac{5}{x}$, then $15x = 45$ and $x = 3$.

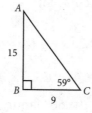 

The **trigonometry method** says if you have a right triangle with an acute angle of 59°, you know that the side opposite the 59° angle is 5 and you are looking for the adjacent side, you can write $\tan(59°) = \dfrac{\text{opposite}}{\text{adjacent}} = \dfrac{5}{x}$ and you can find out that $\tan(59°)$ is approximately 1.66. That takes the place of the ratio from $\triangle ABC$, and you can check that $15 \div 9$ is very close to the value the calculator gives for $\tan(59°)$.

$$\tan(59°) = \frac{5}{x}$$
$$1.66 \approx \frac{5}{x}$$
$$1.66x \approx 5$$
$$x \approx \frac{5}{1.66} \approx 3.01$$

# Finding a Side

Because you're able to call up the value of a particular ratio for any family, you can find the length of an unknown side without needing to have two similar triangles. You just need to have one right triangle, know the measurement of one acute angle, and know the length of one side.

To use trig ratios to find a missing side in a right triangle:

Do you know the measure of one acute angle?

↓

Do you know the length of one side?

↓

Find a ratio that uses the side you know and the side you're looking for.

↓

Set up the ratio.

↓

Get the value of that ratio for the family of triangles
with your angle from your calculator or a table.

↓

Solve the equation.

## EXERCISE 10.3

**Use the right triangle below and the definitions of the trig ratios to find the ratios in questions 1 through 6.**

1.  Use a calculator to find each of the following to the nearest thousandth.

    a)  $\sin(78°)$

    b)  $\cos(15°)$

    c)  $\cos(43°)$

    d)  $\tan(62°)$

    e)  $\sin(21°)$

2.  Right triangle $\triangle XYZ$ has hypotenuse $\overline{XZ}$, which measures 38 cm. If $m\angle X = 30°$, find the length of leg $\overline{XY}$ and the length of $\overline{YZ}$.

3.  In right triangle $\triangle ABC$, $\angle C$ measures 79° and side $\overline{AB}$ measures 35 cm. Find the

length of hypotenuse $\overline{AC}$ to the nearest tenth.

4.  Right triangle $\triangle RST$ has a 45° angle and a leg 12 inches long. Find the length of the hypotenuse, to the nearest tenth of an inch.

5.  $\triangle XYZ$ is a right triangle with $m\angle X = 40°$ and one leg $\overline{XY} = 16$ in. Find the length of the other leg, $\overline{YZ}$.

6.  Rectangle $ABCD$ has a diagonal $BD = 12$ inches, and $\angle BDA$ measures 15°. Find the length of $\overline{AD}$.

7.  How long is diagonal $\overline{XZ}$ in rectangle $WXYZ$ if $m\angle WXZ = 47°$ and $XW = 5$ meters?

# Finding an Angle

Trig ratios give you a quick way to find an unknown side of a right triangle, as long as you know the measure of one acute angle of the right triangle. But what if you don't know the measure of the angle? What if the measure of the angle is exactly what you're trying to find?

You can do that with a little more information. You'll need to know two sides of the right triangle, and you'll need the inverse operation. You've met other inverses; square roots undo the work of squaring, for example. When you write $\cos(60) = \frac{1}{2}$, you're saying the right triangles in the 60° family have a ratio of adjacent side to hypotenuse that reduces to $\frac{1}{2}$. What you need now is to say "what angle has a cosine value of $\frac{1}{2}$?" You can write that as $\cos^{-1}\left(\frac{1}{2}\right) = 60°$. The $-1$ is not an exponent but a symbol for the inverse or opposite process. To find the angle that has a sine equal to $\frac{\sqrt{2}}{2}$, you want $\sin^{-1}\left(\frac{\sqrt{2}}{2}\right) = 45°$.

There is another notation for the inverse of the trig relationships that is sometimes used. Instead of $\sin^{-1}\left(\frac{\sqrt{2}}{2}\right)$, the same idea could be expressed as $\arcsin\left(\frac{\sqrt{2}}{2}\right)$, $\cos^{-1}\left(\frac{1}{2}\right)$ could be written as $\arccos\left(\frac{1}{2}\right)$, and $\tan^{-1}(1)$ could be $\arctan(1)$.

To find the measure of an acute angle of a right triangle:

You must know the lengths of two sides of the right triangle.

↓

Create the appropriate ratio with those two sides.

↓

Use the corresponding inverse to find the measure of the angle.

EXAMPLE

$\triangle ABC$ has right angle $\angle B$. Hypotenuse $\overline{AC}$ measures 14 cm and leg $AB$ measures 8 cm. To find the measure of $\angle C$, notice that you have the hypotenuse and the side opposite $\angle C$. The opposite and hypotenuse can create the sine ratio. $\sin(\angle C) = \dfrac{8}{14} = \dfrac{4}{7}$. Simplifying the fraction is not absolutely necessary, so use $\dfrac{8}{14}$ if you prefer. Write $m\angle C = \sin^{-1}\left(\dfrac{4}{7}\right)$, and then find $\sin^{-1}$ above the sin key on your calculator. Type 2$^{nd}$ and then sin to get $\sin^{-1}$. Type 4/7, close the parentheses, and press ENTER.

$m\angle C = \sin^{-1}\left(\dfrac{4}{7}\right) \approx 34.850°$ when rounded to the nearest thousandth.

## EXERCISE 10.4

**Use inverse trig functions to complete these questions. Round answers as indicated.**

1. Find each of the following to the nearest degree.

    a) $\tan^{-1}(2.904)$

    b) $\cos^{-1}(0.777)$

    c) $\sin^{-1}(0.819)$

    d) $\cos^{-1}(0.139)$

    e) $\tan^{-1}(1.000)$

2. In right triangle $\triangle ABC$, leg $\overline{AB}$ measures 39 cm and leg $\overline{BC}$ measures 62 cm. Find the measure of $\angle A$ to the nearest tenth of a degree.

3. The hypotenuse of right triangle $\angle JKL$ is $JL = 23$ inches. If leg $\overline{JK}$ measures 14 inches, what is the measure of $\angle L$ to the nearest tenth of a degree?

4. Find the measure of the smaller acute angle of right triangle $\triangle XYZ$ if the legs have measures of $XY = 35$ cm and $YZ = 40$ cm. Round to the nearest tenth of a degree.

5. Find the measure of the larger acute angle of right triangle $\triangle ABC$ if hypotenuse $AC = 20$ inches and shorter leg $AB = 10$ inches.

6. Find the measures of the acute angles of a right triangle if the hypotenuse is $17\sqrt{2}$ cm and a leg measures 17 cm.

7. A right triangle has sides that are a 3-4-5 Pythagorean triple. Find the measure of the acute angles to the nearest tenth of a degree.

# Reciprocal and Complementary Relationships

When the trig ratios were named and defined, you may have noticed that there were two ratios that involved each pair of sides. The six different ratios are three pairs of reciprocals, and like all reciprocal pairs, their product is 1.

$$\sin(\angle A) = \frac{\text{opposite}}{\text{hypotenuse}} \qquad \csc(\angle A) = \frac{\text{hypotenuse}}{\text{opposite}} \qquad \sin(\angle A) \cdot \csc(\angle A) = 1$$

$$\cos(\angle A) = \frac{\text{adjacent}}{\text{hypotenuse}} \qquad \sec(\angle A) = \frac{\text{hypotenuse}}{\text{adjacent}} \qquad \cos(\angle A) \cdot \sec(\angle A) = 1$$

$$\tan(\angle A) = \frac{\text{opposite}}{\text{adjacent}} \qquad \cot(\angle A) = \frac{\text{adjacent}}{\text{opposite}} \qquad \tan(\angle A) \cdot \cot(\angle A) = 1$$

The reciprocals give you a choice when you're trying to find an unknown side. If you know the opposite and are looking for the adjacent, you can use

$$\tan(\angle A) = \frac{\text{opposite}}{\text{adjacent}} = \frac{BC}{AB} \quad \text{or} \quad \cot(\angle A) = \frac{\text{adjacent}}{\text{opposite}} = \frac{AB}{BC}, \text{whichever you}$$

prefer. Each has an advantage. If you are looking for *AB*, you may prefer to have the variable in the numerator, because it's faster to solve. On the other hand, using sin, cos, or tan means you have a key on your calculator that will give you that value. You'll have to work a bit to get the value of sec, csc, or cot.

EXAMPLE

▶ To find the value of csc(37°) on your calculator, use $1 \div \sin(37°)$. To find sec(49°), enter $1 \div \cos(49°)$. If you need cot(65°), you want $1 \div \tan(65°)$.

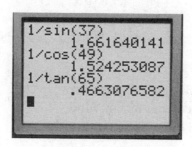

The "co" in the names of three of the six trig ratios comes from "complementary." Two angles are complementary if their measures add to 90°, and the two acute angles in a right triangle are complementary. You may have noticed that the sine of one angle in a right triangle is the same value as the cosine of the other angle. The sine of one angle is the cosine of its complement. The complementary relationships are sine and cosine, secant and cosecant, and tangent and cotangent.

$$\sin(\angle A) = \cos(90° - \angle A) \qquad \cos(\angle A) = \sin(90° - \angle A)$$
$$\sec(\angle A) = \csc(90° - \angle A) \qquad \csc(\angle A) = \sec(90° - \angle A)$$
$$\tan(\angle A) = \cot(90° - \angle A) \qquad \cot(\angle A) = \tan(90° - \angle A)$$

If you're trying to find the measure of an acute angle in $\triangle ABC$, and you know the lengths of leg $AB$ and hypotenuse $AC$, you can use

$$\cos(\angle A) = \frac{\text{adjacent}}{\text{hypotenuse}} = \frac{AB}{AC} \text{ or } \sin(\angle C) = \frac{\text{opposite}}{\text{hypotenuse}} = \frac{AB}{AC}, \text{ depending}$$

on which angle you want to find. The cosine of $\angle A$ is the sine of $\angle C$.

## EXERCISE 10.5

**In this set of questions, use your calculator when necessary for arithmetic, but do not use it to find trig ratios. Instead, deduce the values of trig ratios from the information given and the rules in the lesson.**

1. If $\sin(\angle A) = \frac{3}{8}$, what is $\csc(\angle A)$?

2. If $\sec(\angle R) = \frac{12}{7}$, what is $\cos(\angle R)$?

3. If $\tan(\angle X) = 3.1$, what is $\cot(\angle X)$?

4. If you know the values of $\sin(\angle L)$ and $\cos(\angle L)$, explain how you could find $\tan(\angle L)$.

5. If $\cos(60°) = \frac{1}{2}$ and $\sin(60°) = \frac{1}{2}\sqrt{3}$, what is $\tan(60°)$?

6. If $\tan(45°) = 1$ and $\sin(45°) = \frac{\sqrt{2}}{2}$, what is $\cos(45°)$?

7. If $\cot(40°) \approx 1.192$ and $\sin(40°) = 0.643$, find $\cos(40°)$ to the nearest thousandth.

8. If $\tan(75°) = 3.732$, what is $\cot(15°)$?

9. If $\cos(60°) = 0.5$, what is $\sin(30°)$?

10. If $\tan(75°) = 3.732$, what is $\cot(75°)$?

11. If $\triangle ABC$ is a right triangle with right angle $\angle B$ and $\sin(\angle A) = 0.3$, find the value, to the nearest thousandth, of:

   a) $\cos(\angle A)$

   b) $\tan(\angle A)$

   c) $\cot(\angle A)$

   d) $\sec(\angle A)$

   e) $\csc(\angle A)$

   f) $\sin(\angle C)$

   g) $\cos(\angle C)$

   h) $\tan(\angle C)$

   i) $\cot(\angle C)$

   j) $\sec(\angle C)$

   k) $\csc(\angle C)$

# Pythagorean Identity

The famous rule known as the Pythagorean theorem says that if $a$ and $b$ are the lengths of the legs of a right triangle, and $c$ is the length of the hypotenuse, then $a^2 + b^2 = c^2$. All of the trig ratios involve the lengths of sides of right triangles, and the Pythagorean theorem provides some information about the relationships among the ratios. Consider right triangle $\triangle RST$ with right angle $\angle S$, and label legs $\overline{RS}$ and $\overline{ST}$ as $a$ and $b$, and hypotenuse $\overline{RT}$ as $c$. Think about the trig ratios for $\angle T$.

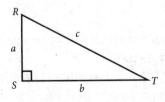

$$\sin(\angle T) = \frac{RS}{RT} = \frac{a}{c} \qquad \cos(\angle T) = \frac{ST}{RT} = \frac{b}{c}$$

Square each ratio by squaring the numerator and squaring the denominator.

$$\sin^2(\angle T) = \frac{a^2}{c^2} \qquad \cos^2(\angle T) = \frac{b^2}{c^2}$$

They have a common denominator, so look at what happens when they're added.

$$\sin^2(\angle T) + \cos^2(\angle T) = \frac{a^2}{c^2} + \frac{b^2}{c^2} = \frac{a^2 + b^2}{c^2}$$

The Pythagorean theorem says $a^2 + b^2 = c^2$.

$$\sin^2(\angle T) + \cos^2(\angle T) = \frac{a^2}{c^2} + \frac{b^2}{c^2} = \frac{a^2 + b^2}{c^2} = \frac{c^2}{c^2} = 1$$

$$\sin^2(\angle T) + \cos^2(\angle T) = 1$$

For any acute angle, the square of the sine of the angle plus the square of the cosine of the angle always equals one.

This observation, known as the **Pythagorean identity**, allows you to find the cosine of an angle if you know its sine, or find the sine if you know the cosine.

---

**EXAMPLE**

If you know that $\sin(30°) = \frac{1}{2}$, then:

$$\sin^2(30°) + \cos^2(30°) = 1$$

$$\left(\frac{1}{2}\right)^2 + \cos^2(30°) = 1$$

$$\frac{1}{4} + \cos^2(30°) = 1$$

$$\cos^2(30°) = 1 - \frac{1}{4} = \frac{3}{4}$$

$$\cos(30°) = \pm\sqrt{\frac{3}{4}} = \pm\frac{\sqrt{3}}{\sqrt{4}} = \pm\frac{\sqrt{3}}{2}$$

---

For now, only the positive square root will make sense in your work, but later the negative option will play a role.

While $\sin^2(\angle A) + \cos^2(\angle A) = 1$ is what is usually thought of as the Pythagorean identity, it is actually one of three. Divide each term by $\sin^2(\angle A)$ to get an identity involving cotangent and cosecant.

$$\frac{\sin^2(\angle A)}{\sin^2(\angle A)} + \frac{\cos^2(\angle A)}{\sin^2(\angle A)} = \frac{1}{\sin^2(\angle A)}$$
$$1 + \cot^2(\angle A) = \csc^2(\angle A)$$

Or divide the original by $\cos^2(\angle A)$ to get an identity involving tangent and secant.

$$\frac{\sin^2(\angle A)}{\cos^2(\angle A)} + \frac{\cos^2(\angle A)}{\cos^2(\angle A)} = \frac{1}{\cos^2(\angle A)}$$
$$\tan^2(\angle A) + 1 = \sec^2(\angle A)$$

## EXERCISE 10.6

**Answer all questions by using the relationships described in this lesson. You may use your calculator for arithmetic.**

**1.** If $\sin(18°) = 0.309$, find $\cos(18°)$.

**2.** If $\cos(27°) = 0.891$, find $\sin(27°)$.

**3.** If $\tan(35°) = 0.700$, find $\sec(35°)$.

**4.** If $\csc(42°) = 1.494$, find $\cot(42°)$.

**5.** If $\tan(59°) = 1.664$, find $\cos(59°)$.

**6.** If $\cos(61°) = 0.485$, find $\tan(61°)$.

**7.** If $\tan(75°) = 3.732$, find $\sin(75°)$.

# Modeling with Trigonometry

Right triangle trigonometry can often provide tools for modeling real-life situations and solving problems. When those situations are described, the description often includes "angle of elevation" or "angle of depression." These angles provide important information but may or may not actually be in the right triangle.

EXAMPLE

Imagine you are standing on the field at a stadium. You look straight ahead to the goal post at the end of the field. Then you notice a flag flying over the stadium. You raise your eyes to look up at the flag. The angle of elevation is the angle between your line of sight straight ahead and your line of sight up to the flag. Depending on what you're looking for, the angle of elevation may be an acute angle of your right triangle.

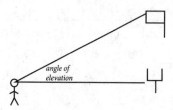

*angle of elevation*

On the other hand, suppose you are standing on the roof of an office building, looking straight ahead at the city skyline. You realize you can see the park and a fountain in the park that you particularly like. You shift your gaze down to the fountain. The angle of depression is the angle between your straight-ahead sight line and your sight line to the fountain. The angle of depression may not seem to be an angle of the right triangle formed by the ground, the building and your line of sight, but your straight-ahead line is parallel to the ground, so there is an angle by the fountain that is equal to the angle of depression.

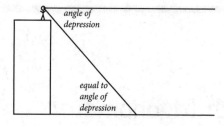

*angle of depression*

*equal to angle of depression*

## EXERCISE 10.7

**For each of the questions in this section, sketch a figure and label it with the given information. Then write and solve an equation to find the requested information.**

1. From a point 250 yards from the foot of a building, the angle of elevation to the top of the building is 35°. How tall is the building to the nearest foot? (3 feet = 1 yard)

2. The angle of depression from the top of a security tower to the nearby building entrance is 42°. If the tower is 25 feet high, how far is it from the entrance to the base of the tower to the nearest tenth of a foot?

3. To the nearest degree, what angle does a stairway make with the floor if the steps have a tread of 9 inches and a rise of 7 inches?

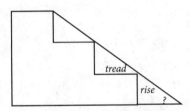

4. From onboard a ship at sea, the angle of elevation to the top of a lighthouse on the shore is 7°. If the lighthouse is known to be 165 feet high, how far from shore is the ship?

5. A ladder 25 feet long makes an angle of 12° with the wall of a building. How far from the wall is the foot of the ladder?

6. From the top of a ski slope, Marta sees the lodge at the base of the slope, at an angle of depression of 20°. If signs by the ski trail tell Marta she is now at an elevation of 1,500 feet, how far will Marta have to ski down the slope to reach the lodge?

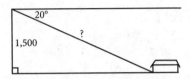

7. From a point 100 feet from the base of the schoolhouse, the angle of elevation to the bottom of a flagpole on the roof of the school house is 40°. Find the height of the schoolhouse.

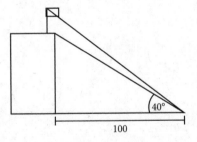

8. If the angle of elevation to the top of the flagpole in question 7 is 42.6°, how tall is the flagpole?

## CHAPTER 11

# Trigonometry in Other Triangles

Right triangle trigonometry provides some powerful tools for finding the length of sides or the measure of angles in right triangles. Unfortunately, not every situation in which information is needed involves a right triangle. Finding ways to apply trigonometric ideas to other triangles is an important next step.

## Law of Cosines

Consider $\triangle ABC$ below, which is not a right triangle. One simple way to create right triangles is to draw an altitude $\overline{BD} \perp \overline{AC}$. To make it a little easier to talk about the lengths of sides, label $\overline{BC}$, the side opposite $\angle A$, as $a$, the side opposite $\angle B$ as $b$, and the side opposite $\angle C$ as $c$. Label the altitude $h$. The altitude divides side $b$ into two sections, so label $DC$ as $x$ and $AD$ as $b - x$.

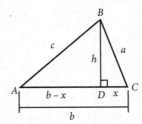

The altitude has created not one but two right triangles, and if the Pythagorean theorem is applied to each of them, that produces two equations: $a^2 = x^2 + h^2$ and $c^2 = (b - x)^2 + h^2$. Use the first equation to say $a^2 - x^2 = h^2$ and substitute into the second equation: $c^2 = (b - x)^2 + a^2 - x^2$. Simplify that equation.

**145**

$$c^2 = b^2 - 2bx + x^2 + a^2 - x^2$$
$$c^2 = a^2 + b^2 - 2bx + x^2 - x^2$$
$$c^2 = a^2 + b^2 - 2bx$$

What else can we say about $x$? $\cos(\angle C) = \dfrac{x}{a}$ so $x = a\cos(\angle C)$. Plug that in.

$$c^2 = a^2 + b^2 - 2bx$$
$$c^2 = a^2 + b^2 - 2b\big(a\cos(\angle C)\big)$$
$$c^2 = a^2 + b^2 - 2ab\cos(\angle C)$$

This equation, which looks like a modified Pythagorean theorem, is called the **Law of Cosines**, and allows you to find an unknown side or an unknown angle of a triangle that is not a right triangle. Of course, you need to have the right information first, and you can remember what's needed by borrowing the abbreviations for proving triangles congruent.

▶ **SAS**   If you know the lengths of two sides of the triangle and you know the measure of the angle included between them, you can use the Law of Cosines to find the length of the third side.

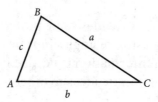

▶ If in the triangle above, $a = 21$ cm, $b = 23$ cm, and $m\angle C = 20°$, the Law of Cosines will give the value of $c$.

$$c^2 = a^2 + b^2 - 2ab\cos(\angle C)$$
$$c^2 = (21)^2 + (23)^2 - 2(21)(23)\cos(20°)$$
$$c^2 = 441 + 529 - 966\cos(20°)$$
$$c^2 = 970 - 966(0.9396)$$
$$c^2 = 970 - 907.7431$$
$$c^2 = 62.2569$$
$$c = \sqrt{62.2569} \approx 7.890$$

▶ Be careful when evaluating to observe the order of operations. Multiply $2ab$ by $\cos(\angle C)$ before doing the subtraction. Taking the square root of both sides theoretically produces both a positive and a negative value, but in this situation only the positive value makes sense.

The Law of Cosines could be rewritten, changing the places of the sides, for example $a^2 = b^2 + c^2 - 2bc\cos(\angle A)$. Just remember that the side whose square is alone on the left must be the side opposite the angle whose cosine is taken.

$$a^2 = b^2 + c^2 - 2bc\cos(\angle A)$$

$$b^2 = a^2 + c^2 - 2ac\cos(\angle B)$$

$$c^2 = a^2 + b^2 - 2ab\cos(\angle C)$$

▶ **SSS**   If you know the lengths of all three sides of the triangle, you can use the Law of Cosines to find the measure of an angle.

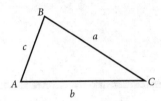

**EXAMPLE**

▶ Suppose that $a = 18$, $b = 22$, and $c = 14$, but no angles are known. Let's find the measure of $\angle A$. That means we want to use $\cos(\angle A)$, so we need to start with $a^2$.

$$a^2 = b^2 + c^2 - 2bc\cos(\angle A)$$
$$(18)^2 = (22)^2 + (14)^2 - 2(22)(14)\cos(\angle A)$$

Solving this version will be a little more complicated, so pay particular attention to order of operations.

$$(18)^2 = (22)^2 + (14)^2 - 2(22)(14)\cos(\angle A)$$
$$324 = 484 + 196 - 616\cos(\angle A)$$
$$324 = 484 + 196 - 616\cos(\angle A)$$
$$324 = 680 - 616\cos(\angle A)$$
$$-356 = -616\cos(\angle A)$$
$$0.5779 = \cos(\angle A)$$
$$\angle A = \cos^{-1}(0.5779)$$
$$\angle A \approx 54.697°$$

## EXERCISE 11.1

**Use the Law of Cosines to solve these questions. Round side lengths to the nearest tenth and angle measures to the nearest degree.**

**1.** In $\triangle ABC$, $AB = 14$ cm, $BC = 18$ cm, and $m\angle B = 32°$. Find the length of $\overline{AC}$.

**2.** In a different $\triangle ABC$, $BC = 29$ inches, $AC = 31$ inches, and $m\angle C = 65°$. Find the length of $\overline{AB}$.

**3.** In $\triangle XYZ$, $YZ = 21$ feet, $XZ = 17$ feet, and $m\angle Z = 54°$. Find the length of $\overline{XY}$.

**4.** In $\triangle RST$, $RT = 33$ cm, $ST = 44$ cm, and $m\angle T = 49°$. Find the length of $\overline{RS}$.

**5.** In $\triangle ABC$, $AB = 14$ inches, $BC = 19$ inches, and $AC = 22$ inches. Find the measure of $\angle B$.

**6.** In the same $\triangle ABC$ in exercise 5, find the measure of $\angle A$.

**7.** In $\triangle XYZ$, $XY = 35$ feet, $YZ = 48$ feet, and $XZ = 42$ feet. Find the measure of $\angle X$.

## Law of Sines

Let's return to that triangle in which we drew an altitude. We applied the Pythagorean theorem last time, but let's look at trig ratios this time.

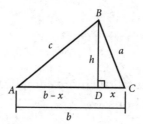

Using the two right triangles the altitude created, $\sin(\angle A) = \dfrac{h}{c}$ and $\sin(\angle C) = \dfrac{h}{a}$. With a little algebra, that becomes $h = c \sin(\angle A)$ and $h = a \sin(\angle C)$, and since both are $h$, $c \sin(\angle A) = a \sin(\angle C)$. Divide both sides by $\sin(\angle A)\sin(\angle C)$:

$$c\sin(\angle A) = a\sin(\angle C)$$

$$\frac{c\,\sin(\angle A)}{\sin(\angle A)\sin(\angle C)} = \frac{a\,\sin(\angle C)}{\sin(\angle A)\sin(\angle C)}$$

$$\frac{c}{\sin\angle C} = \frac{a}{\sin\angle A}$$

You can probably imagine that by drawing a different altitude, you could derive the same relationship involving $a$ and $b$ or for $b$ and $c$. So the **Law of**

**Sines** is summarized as $\dfrac{a}{\sin\angle A} = \dfrac{b}{\sin\angle B} = \dfrac{c}{\sin\angle C}$ and you get to choose which two ratios work for your problem. All rules for proportions hold, so if you'd prefer $\dfrac{\sin\angle A}{a} = \dfrac{\sin\angle B}{b}$ or $\dfrac{b}{c} = \dfrac{\sin\angle B}{\sin\angle C}$, you can do that.

Congruent triangle abbreviations help here too. There is no SSS case, because you need to know at least one angle or you won't be able to solve any proportion.

▶ **ASA** and **AAS**   If you know the measures of two angles of the triangle, you can find the third, so you don't need to worry about whether your given information is two angles and the included side or two angles and a non-included side. Knowing two angles and a side allows you to find another side.

EXAMPLE

▶ Suppose m$\angle A = 60°$, m$\angle C = 25°$, and $a = 32$. You can determine that m$\angle B = 180 - (60 + 25) = 95°$.

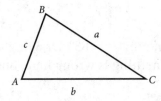

▶ Plug in what you know. $\dfrac{a}{\sin\angle A} = \dfrac{b}{\sin\angle B} = \dfrac{c}{\sin\angle C}$ becomes $\dfrac{32}{\sin(60°)} = \dfrac{b}{\sin(95°)} = \dfrac{c}{\sin(25°)}$. You'll want to use the first ratio, in which you know both numerator and denominator, and then choose another based on whether you want to find $b$ or $c$. Let's find $c$.

$$\frac{32}{\sin(60°)} = \frac{c}{\sin(25°)}$$
$$c \cdot \sin(60°) = 32\sin(25°)$$
$$c = \frac{32\sin(25°)}{\sin(60°)}$$
$$c \approx \frac{32(0.4226)}{0.8660} \approx 15.616$$

You can put in the decimal approximations for the sines earlier in the solution if you wish, but they are usually decimals that you'll have to round to three or four places, many people find it simpler to wait until the end and do all the calculator work at once.

▶ **SSA**  Yes, you're correct. SSA is not a valid way to prove triangles congruent. That, in fact, makes it a perfect mnemonic for this case. When you know two sides and one angle, you may or may not be successful in finding another angle. If you had SAS, two sides and the included angle, you won't have sufficient information to solve a proportion. For example, if you knew that in $\triangle XYZ$, $XY = 5$ inches, $YZ = 9$ inches, and $m\angle Y = 52°$, and you tried to set up a proportion, you'd have

$$\frac{YZ}{\sin\angle X} = \frac{XZ}{\sin\angle Y} = \frac{XY}{\sin\angle Z}$$, and you'd substitute the measurements you knew. That would give you $\frac{9}{\sin\angle X} = \frac{XZ}{\sin(52°)} = \frac{5}{\sin\angle Z}$. There's no way to choose two ratios without having two unknowns.

If you have SSA, you may get a solution, you may get two, or you may find out that there is no triangle that fits. Let's look at the last case first. Remember that sine is defined as the ratio of opposite to hypotenuse. Because the hypotenuse is the longest side of a right triangle, sine will never be larger than one.

EXAMPLE

▶ Suppose $a = 8$, $b = 12$, and $m\angle A = 85°$. Drawing a picture may be enough for you to see something is wrong here, but let's do the math.

$$\frac{a}{\sin\angle A} = \frac{b}{\sin\angle B}$$

$$\frac{8}{\sin(85°)} = \frac{12}{\sin\angle B}$$

$$8\sin\angle B = 12\sin(85°)$$

$$\sin\angle B = \frac{12\sin(85°)}{8} \approx \frac{12(0.9962)}{8} \approx 1.4943$$

▶ There is no angle that has a sine greater than 1, so we have to conclude that there is no solution.

▶ Because these rules are for triangles that are not right triangles, the angles may be acute or obtuse, and one of the complications is that acute angles and their supplements, which are obtuse, have the same sine. For example, $\sin(30°) = \frac{1}{2}$ but $\sin(150°) = \frac{1}{2}$ as well. The $\sin^{-1}($ key on your calculator will only give you the acute angle. You have to think about whether the obtuse angle might also work.

EXAMPLE

Suppose $a = 12$, $b = 8$, and $m\angle A = 50°$. Set up and solve. Round to the nearest whole degree.

$$\frac{12}{\sin 50°} = \frac{8}{\sin\angle B}$$
$$12\sin\angle B = 8\sin 50°$$
$$\sin\angle B = \frac{8\sin 50°}{12} \approx 0.5107$$
$$\sin^{-1}(0.5107) \approx 31°$$

▶ That solution means the triangle has angles of 50°, not quite 31°, and 180° − (50° + 31°) = 99°. But now you need to think about the supplement of 31°, an obtuse angle of 149°. Is it possible to have a triangle with angles of 50° and 149° and a third angle? 50° + 149° is already 199°, and triangles only have 180°. This obtuse triangle option won't work, so you have one solution: angles of 50°, 31°, and 99°.

▶ Now consider $a = 8$, $b = 12$, and $m\angle A = 40°$.

$$\frac{8}{\sin 40°} = \frac{12}{\sin\angle B}$$
$$8\sin\angle B = 12\sin 40°$$
$$\sin\angle B = \frac{12\sin 40°}{8} \approx 0.9642$$
$$\sin^{-1}(0.9642) \approx 75°$$

▶ This could be an acute triangle with angles of 40°, 75°, and 180° − (40° + 75°) = 65°.

▶ Could it be an obtuse triangle with angles of 40°, 180° − 75° = 105°, and 180° − (40° + 105°) = 35°? That is a possibility. Angles of 40°, 105°, and 35° could make a triangle. So, here, you have to conclude that there are two possible triangles, one acute and one obtuse.

## EXERCISE 11.2

**Use the Law of Sines to solve these problems. In these exercises, round side lengths to the nearest tenth and angle measures to the nearest degree. If more than one solution is possible, give both. If no solution is possible, so state.**

1.  In $\triangle ABC$, $m\angle A = 73°$, $m\angle B = 42°$, and $BC = 15$ cm. Find the length of $\overline{AC}$.

2.  In a different $\triangle ABC$, $m\angle B = 100°$, $m\angle C = 19°$, and $AC = 38$ inches. Find the length of $\overline{AB}$.

3.  In $\triangle RST$, $m\angle R = 53°$, $m\angle T = 62°$, and $RS = 120$ feet. Find the length of $\overline{ST}$.

4.  In $\triangle XYZ$, $m\angle Y = 93°$, $m\angle Z = 45°$, and $XZ = 85$ m. Find the length of $\overline{XY}$.

**5.** In $\triangle ABC$, $m\angle B = 55°$, $AC = 22$ inches, and $AB = 34$ inches. Find the measure of $\angle C$.

**6.** In $\triangle XYZ$, $m\angle Z = 98°$, $XY = 71$ cm, and $YZ = 63$ cm. Find the measure of $\angle X$.

**7.** In $\triangle RST$, $m\angle R = 14°$, $ST = 83$ feet, and $RT = 125$ feet. Find the measure of $\angle S$.

# Solving Triangles

The expression "solve the triangle" is simply a quick way to say "find all sides and all angles not already known." To do this, you can use the Law of Cosines or the Law of Sines, or both, in any convenient order. It simply depends on what information you have, and what you are trying to find.

EXAMPLE

Suppose $\triangle ABC$ has $a = 18$, $b = 36$, and $m\angle C = 147°$. Solve the triangle, rounding to the nearest whole number when necessary.

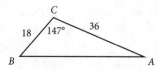

Start with the Law of Cosines because the given information is SAS.

$$c^2 = a^2 + b^2 - 2ab\cos(\angle C)$$
$$c^2 = 18^2 + 36^2 - 2(18)(36)\cos(147°)$$
$$c^2 = 324 + 1296 - 1296(-0.8387)$$
$$c^2 = 1620 + 1086.9552$$
$$c^2 = 2706.9552$$
$$c = \sqrt{2706.9552} \approx 52.0284$$

Side $c \approx 52$. All three sides are known. Let's use Law of Sines to find another angle. Don't forget to check if two triangles are possible.

$$\frac{a}{\sin(\angle A)} = \frac{c}{\sin(\angle C)}$$
$$\frac{18}{\sin(\angle A)} = \frac{52}{\sin(147°)}$$
$$52\sin(\angle A) = 18\sin(147°)$$
$$\sin(\angle A) = \frac{18\sin(147°)}{52} \approx \frac{18(0.5446)}{52} \approx 0.1885$$
$$\angle A = \sin^{-1}(0.1885) \approx 10.867$$

$\angle A \approx 11°$, so the three angles are 147°, 11°, and 23°. Is it possible to use the supplement of 11°? That would be 169°, which can't possibly be in the same triangle with an angle of 147°. There is one solution.

## EXERCISE 11.3

**Some of these questions will require the Law of Sines, some the Law of Cosines, some both. Choose the most convenient method to find each unknown measurement. In these exercises, round side lengths to the nearest tenth and angle measures to the nearest degree. If more than one solution is possible, give both solutions in full.**

1. Find the measures of each of the angles of a triangle with sides of 6 cm, 11 cm, and 14 cm.

2. In $\triangle ABC$, m$\angle A = 62°$, m$\angle B = 88°$, and $AB = 51$ inches. Find the measurements of the remaining sides and angles.

3. Solve $\triangle XYZ$ if $XY = 47$ feet, $YZ = 58$ feet, and m$\angle Y = 81°$.

4. Solve $\triangle RST$ if $RS = 19$ m, $ST = 25$ m, and m$\angle T = 43°$.

5. Mr. Johnson's home has a porch that is 25 ft long. Mr. Johnson wants to construct a triangular garden with that porch as one side, and he has marked off an angle of 80° at one end of the porch, and an angle of 65° at the other end. Find the lengths of the remaining sides of the garden.

6. A small park in the shape of a triangle has sides of 200 feet, 217 feet, and 226 feet. Find the angles at the corners of the park.

7. Chris is designing a triangular sail. One side of the sail must be 24 feet long, and another side 36 feet. Those sides meet at an angle of 92°. Find the remaining side and angles.

# Area of a Triangle

The area of a right triangle is equal to half the product of the lengths of the legs. If you know the sides of the right triangle, you can find its area. For other triangles, however, knowing all three sides is not enough. The area formula calls for the height, or length of an altitude. With the help of trigonometry, we can derive a variant area formula that relies on sides and the measure of one angle.

EXAMPLE

▶ In the triangle below, if the lengths of $h$ and $b$ were known, the area of the triangle could be calculated as $A = \frac{1}{2}bh$. If $h$ is not known, we may be able to find it, if we have side $a$ and $\angle C$ as well as side $b$. We know $\sin(\angle C) = \frac{h}{a}$, so $h = a\sin(\angle C)$. Then the area $A = \frac{1}{2}bh$ becomes $A = \frac{1}{2}b(a\sin\angle C) = \frac{1}{2}ab\sin\angle C$.

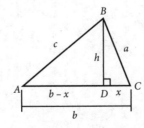

▶ This area formula can be adapted, as long as you know two sides and the included angle. If you know sides $a$ and $c$ and $\angle B$, your formula is $A = \frac{1}{2}ac\sin\angle B$.

## EXERCISE 11.4

**Use the area formulas introduced in this section to solve each problem. In these exercises, round side lengths to the nearest tenth and angle measures to the nearest degree. Round areas to the nearest tenth of a square unit.**

1. Find the area of a triangle with sides of 15 inches and 18 inches, if the included angle measures 49°.

2. If $\triangle ABC$ has $m\angle C = 12°$, $AC = 27$ cm, and $BC = 31$ cm, what is the area of the triangle?

3. In $\triangle XYZ$, $XY = 44$ feet, $YZ = 50$ feet, and $m\angle Y = 107°$. What is the area of $\triangle XYZ$?

4. In $\triangle RST$, $RS = 112$ m, $ST = 125$ m, and $m\angle S = 93°$. Find the area of $\triangle RST$.

5. A triangle with an area of 413.3 square inches has sides of 24 inches and 38 inches. What is the measure of the angle included between those two sides?

6. In $\triangle ABC$, $BC = 40$ cm and $m\angle C = 82°$. If the area of $\triangle ABC$ is 1,287.3 cm², find the length of $AC$.

7. $\triangle RST$ has an area of 190.7 m². If $ST = 35$ m and $m\angle T = 27°$, how long is $RT$?

# Similarity, Right Triangle Trigonometry, and Trigonometry in Other Triangles

The questions in this review section brings together the concepts from Chapters 9, 10, and 11. Use this opportunity to test your understanding of similar triangles and the application of those ideas to solve for missing measurements by using trigonometry in triangles, whether right, acute, or obtuse. Answer all the questions, and try to express your thinking as clearly as you can.

1. Write a correct similarity statement for the triangles below.

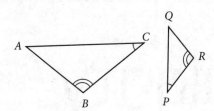

2. If there is sufficient information to conclude that the triangles below are similar, write an extended proportion relating the sides. If there is not enough information, tell what else is needed.

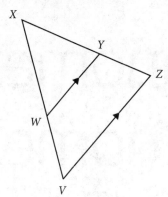

3. In the figure below, $\overline{AB} \parallel \overline{ED}$. Prove $\dfrac{BC}{CD} = \dfrac{AC}{CE}$.

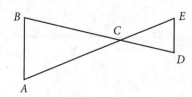

4. In the figure below, $\angle X \cong \angle ZWY$. Prove $\triangle WYZ \sim \triangle XWZ$.

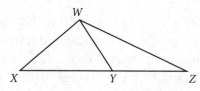

5. In $\triangle ABC$, $\angle DEA \cong \angle EAC$. Prove $\dfrac{BD}{DA} = \dfrac{BE}{EC}$.

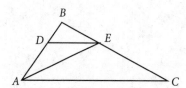

**6.** In $\triangle PQR$, $\overline{RS} \cong \overline{SQ}$ and $\angle PRS \cong \angle Q$. Prove $\dfrac{PS}{PR} = \dfrac{SQ}{RQ}$.

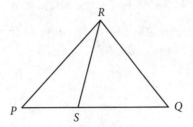

**7.** In $\triangle JKL$, $JK = 17$ cm and $JL = 25$ cm. If $\triangle JKL \sim \triangle DEF$ and $DF = 30$ cm, find $DE$.

**8.** $\triangle ABC \sim \triangle XYZ$. If $AB = x - 3$ cm, $BC = 3$ cm, $AC = x - 1$ cm, $YZ = 21$ cm, and $XY = 2x + 4$ cm, find the perimeter of $\triangle XYZ$.

**9.** At a certain time of day, a 10-foot tree that stands perpendicular to the ground casts a shadow on the ground. The tip of the shadow is 18 feet from the base of the tree. How far from the base of the tree should a 6-foot tall man stand so that the tip of his shadow falls at the tip of the shadow of the tree?

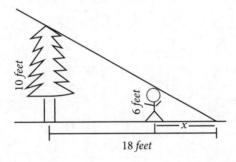

**10.** Jason and Brianna are constructing a scale model of a park in their neighborhood. The park is triangular and their model is similar to the original, but reduced in size. If one side of the park measures 153 meters, and is represented on the model by an edge 38.25 cm long, how long should Jason and Brianna make the edge representing the 119-meter side?

**11.** Use the triangle below to express each of the following.

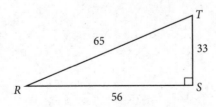

a) $\sin \angle R$

b) $\tan \angle T$

c) $\cos \angle R$

d) $\cos \angle T$

e) $\csc \angle R$

12. In right triangle $\triangle ABC$, $m\angle B = 90°$, $m\angle C = 34°$, and $AB = 12$ cm. Find the length of the hypotenuse.

13. Find the measure of the smallest acute angle in a 3-4-5 right triangle.

Use the following information for questions 14 through 16. In right triangle $\triangle RST$ with hypotenuse $\overline{RT}$, $\sin\angle R = 0.2$.

14. Find $\csc\angle R$.

15. Find $\sec\angle T$.

16. Find $\cos\angle R$.

17. In $\triangle JKL$, $m\angle J = 63°$, $m\angle L = 51°$, and $JL = 20$ cm. Find the length of $\overline{KL}$.

18. In $\triangle XYZ$, $XY = 45$ cm, $YZ = 52$ cm, and $XZ = 55$ cm. Find the measure of $\angle X$.

19. Find the area of $\triangle ABC$ if $AB = 18$ inches, $BC = 24$ inches, and $m\angle B = 65°$.

20. The area of $\triangle XYZ$ is $A = 775.9$ square meters. If $XY = 54$ m and $YZ = 68$ m, find the measure of $\angle Y$.

## CHAPTER 12

# Circles

Much of the study of geometry is focused on polygons, figures built from line segments. The primary exception is the circle, a simple figure with so much to explore.

## Circle Vocabulary

A **circle** is the set of all points in the plane at a fixed distance from a fixed point. The fixed point is called the **center**. The fixed distance is called the **radius**. The term *radius* is also used to refer to a line segment that connects the center to a point on the circle. A line segment that has its endpoints on the circle and passes through the center is the **diameter**.

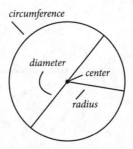

The distance around the circle is called the **circumference**, an idea similar to the perimeter of a polygon. The **area** of a circle is the area enclosed within it. The calculations of circumference and area rely on a constant, π, which is the ratio of the circumference to the diameter. The number π is usually approximated as 3.14 or $\frac{22}{7}$.

A portion of the circle, between two endpoints, is called an **arc**. Arcs are measured in degrees, with the full circle being 360°. Half a circle, called a **semicircle**, is an arc of 180°. An arc that is less than 180°, less than a semicircle, is a **minor arc**, and one that measures more than 180° but less than 360° is a **major arc**.

A circle is named by its center, as in "circle *O*," but minor arcs are named by their endpoints, with an arc symbol above, as in $\overset{\frown}{AB}$. To distinguish a major arc from a minor arc, major arcs are named with three letters, the two endpoints, and another point on the arc between the endpoints, as in $\overset{\frown}{ABC}$, which begins at *A*, passes through *B*, and ends at *C*.

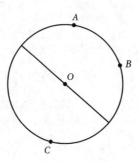

▶ **Chords and Diameters**   In addition to the radius and diameter, there are a number of lines and line segments in and around circles. A line segment whose endpoints are points on the circle is a **chord**. The diameter is a chord that passes through the center of the circle; as such, it is the longest chord in the circle.

▶ **Secants and Tangents**   If a chord is extended beyond the circle, the resulting line is called a **secant**. A **tangent** is a line that touches the circle at only one point. Two circles may have a common tangent. If that tangent passes between the two circles, so that it crosses a line segment connecting the centers, it is an *internal tangent*. If it does not cross that line segment, it is *externally tangent* to the two circles.

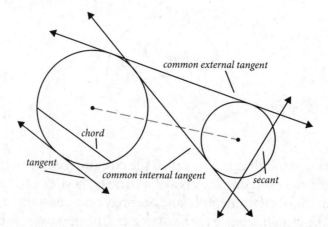

Circles may be tangent to one another, if they touch at just one point. If the two circles bump up against one another, they are externally tangent. If one circle is inside the other, just touching one point on the outer circle, the circles are internally tangent. If one circle sits inside another so that they have the same center and do not touch, they are called **concentric circles**.

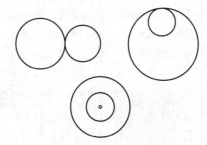

**These questions focus on the vocabulary of circles and on correct notation. Use the figure below for questions 1 through 7.**

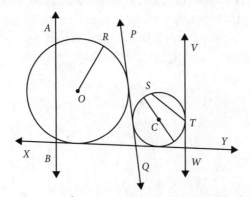

**1.** Name a chord.

**2.** Name a common external tangent.

**3.** Name a secant.

**4.** Name a minor arc.

**5.** Describe $\overline{OR}$.

**6.** Describe $\overleftrightarrow{PQ}$.

**7.** In circle $C$, two segments have been drawn: a _____ and a _____.

# Circumference and Area of Circles

The circumference of a circle is a measurement of the distance around the circle. Early mathematicians noticed that the ratio of the circumference of a circle to its diameter was always a little more than 3. Different cultures made different estimates of that number, but all a little more than 3.

Today we refer to that number as *pi*, using the symbol $\pi$. Mathematicians have calculated thousands of digits of the decimal of $\pi$, but it has no end, so if we want an exact value, we'll just say $\pi$ and if we want an approximate value, we'll use one of the common approximations: 3.14 or $\frac{22}{7}$. Calculators generally have a key for $\pi$, which will give more decimal places, like 3.141592654, but that is still approximate.

▶ **Circumference**   $C = \pi d = 2\pi r$, where $d$ is the diameter and $r$ is the radius. You can use whichever version of the formula is convenient for the information you have and the value for which you're looking.

▶ **Area**    $A = \pi r^2$, where $r$ is the radius. By overlaying a circle with four squares, each with sides the length of the radius, you can see that $4r^2$ is larger than the area of the circle.

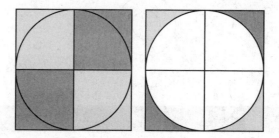

The actual formula can be derived by cutting the circle into sectors and rearranging them to create something that looks a lot like a parallelogram. The smaller you make the sectors, the more of them you have, and the more the rearrangement resembles a parallelogram. That near-parallelogram has a height equal to the radius, and a slightly bumpy base equal to half the circumference, or $\pi r$. The area of the rearranged circle is $A = r(\pi r)$ or $A = \pi r^2$.

## EXERCISE 12.2

**Use the formulas for the circumference and the area of a circle to answer these questions. Round answers as indicated. Be sure to include appropriate units.**

1. Find the circumference, to the nearest tenth of a foot, of a circle with diameter of 54 inches. (12 inches = 1 foot)

2. Find the radius of a circle that has a circumference of $84\pi$ cm.

3. Find the area of a circle with radius 6 inches. Leave your answer in terms of $\pi$.

4. The area of circle $O$ is approximately 907.92 square meters. To the nearest tenth of a meter, what is the radius of circle $O$?

5. If Leslie has 100 m of fencing to enclose a circular play space for the family dog, what is the largest possible area for that space? Round to the nearest square meter.

6. Jesse wondered whether a pizza with a 12-inch diameter for $12 was a better buy than one with a 10-inch diameter for $8. Help Jesse compare by finding the cost per square inch for each pizza.

7. Find a simple formula to find the area of a circle if its circumference is known.

# Angle Measurement

There are many different types of angles that can be drawn in and around circles, and it may seem as those each has its own rule for measurement, but in fact, they can all be organized under four rules, depending on where the vertex of the angle is.

▶ **Vertex at the center**     measure of the angle = measure of the arc

The central angle takes its name from the fact that its vertex is the center of the circle. Its sides are radii, and the number of degrees of arc between the points at which those radii touch the circle is the measure of both the arc and the central angle.

▶ **Vertex on the circle**     measure of the angle = $\frac{1}{2}$ measure of the arc

An inscribed angle is formed by two chords that meet at a point on the circle. The measure of the angle is half the measure of the arc intercepted by those chords.

   An angle formed by a tangent and a chord has its vertex on the circle, at the point of tangency. The tangent and chord actually form a linear pair, and each angle is half the measure of the arc on that side of the chord.

EXAMPLE

▶ In this figure, the measure of:

▶ Central angle $\angle XOY$ = the measure of arc $\overset{\frown}{XY}$.

▶ Inscribed angle $\angle ABC$ is $\frac{1}{2}$ the measure of arc $\overset{\frown}{AC}$.

▶ $\angle QPR = \frac{1}{2}$ the measure of arc $\overset{\frown}{QP}$ and m$\angle QPS = \frac{1}{2}$ m$\overset{\frown}{QCP}$.

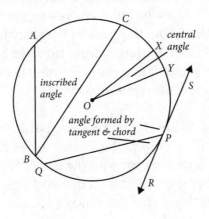

▶ **Vertex inside the circle**     measure of the angle = $\frac{1}{2}$ sum of the two arcs

When two chords intersect within the circle, their endpoints divide the circle into four arcs, and the chords create two pairs of vertical angles. The measure of the angle is the average of the arcs it and its vertical angle partner intercept.

Angle formed by a tangent and a diameter or radius is a special case of tangent and chord, because the diameter is the longest chord, and the radius is a portion of the diameter. The diameter will divide the circle into semicircles so each of these angles will be half of 180°, or a right angle.

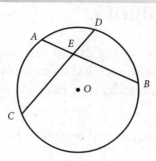

In this figure, $m\angle AED = m\angle CEB = \frac{1}{2}(m\overset{\frown}{AD} + m\overset{\frown}{CB})$ and

$m\angle AEC = m\angle DEB = \frac{1}{2}(m\overset{\frown}{AC} + m\overset{\frown}{DB})$

▶ **Vertex outside the circle**  measure of the angle $= \frac{1}{2}$ difference of the two arcs

▶ **Angle formed by two tangents from the same point**  The tangents each touch the circle at one point. These two points of tangency divide the circle into two arcs: a smaller one near the vertex of the angle, and a larger one that is the remainder of the circle. If the smaller arc is $x°$, the larger one is $360° - x°$, and the measure of the angle is $\frac{1}{2}(360° - x° - x°) = \frac{1}{2}(360° - 2x°) = 180° - x°$

▶ **Angle formed by a tangent and a secant from the same point**  The tangent touches the circle at one point and the secant cuts through two points. The smaller arc is from the point of tangency to the first time the secant intersects the circle. The larger arc is from the second intersection with the secant to the point of tangency. The measure of the angle is $\frac{1}{2}$ (measure of larger arc – measure of smaller arc).

▶ **Angle formed by two secants from the same point**  Each secant intersects the circle in two points. The smaller arc is the arc between the first intersection of each secant with the circle. The larger arc is the portion of the circle between the second intersections. The measure of the angle is $\frac{1}{2}$ (measure of larger arc – measure of smaller arc).

In the following circles:

On the left, $m\angle P = \frac{1}{2}(m\overset{\frown}{ACB} - m\overset{\frown}{AB})$

In the center, $m\angle P = \frac{1}{2}(m\overset{\frown}{RS} - m\overset{\frown}{QS})$

On the right, $m\angle P = \frac{1}{2}(m\overset{\frown}{RT} - m\overset{\frown}{QS})$

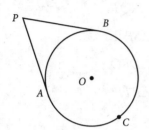 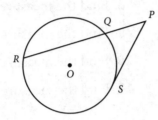 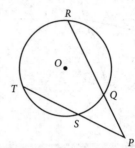

## EXERCISE 12.3

**Use the information on angles in and around circles to answer these questions. Use the figure below for questions 1 through 6.**

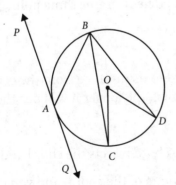

1. If $m\angle DOC = 38°$, what is the measure of $\angle DBC$?

2. If $\overset{\frown}{AB}$ measures 50°, what is the measure of $\angle PAB$?

3. If $m\angle ABC = 21°$, what is the measure of $\overset{\frown}{AC}$?

4. If $\overset{\frown}{AB}$ measures 50°, what is the measure of $\angle QAB$?

5. Based on the information in questions 1 through 4, what is the measure of $\overset{\frown}{BD}$?

6. If $\overline{AD}$ were drawn, what would be the measure of $\angle BAD$?

**Use the figure below for questions 7 through 11.**

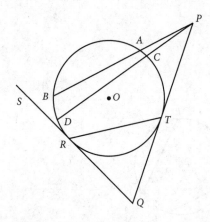

Given that $\overset{\frown}{AB} = 110°$, $\overset{\frown}{BD} = 40°$, $\overset{\frown}{DR} = 30°$, $\overset{\frown}{RT} = 90°$, $\overset{\frown}{TC} = 75°$, $\overset{\frown}{CA} = 15°$.

**7.** Find the measure of $\angle BPD$.

**8.** Find the measure of $\angle RQT$.

**9.** Find the measure of $\angle SRT$.

**10.** Find the measure of $\angle DPQ$.

**11.** Find the measure of $\angle RTP$.

# Segments in Circles

There are several different relationships among segments in and around circles, which aren't quite as easy to categorize. There are the segments that simply are congruent.

▶ In any circle, all radii are congruent, all diameters are congruent, and the length of a diameter is twice the length of a radius.

▶ If two tangent segments are drawn to a circle from the same point, the tangent segments are congruent.

The next two are a little more complicated.

▶ The length of a chord is inversely related to its distance from the center. The closer to the center the chord is drawn, the longer it will be. The longest chord is the diameter.

▶ A diameter or radius perpendicular to a chord bisects the chord and its arc.

Draw radii from the center to the endpoints of the chord, and you can prove the right triangles congruent. The two sections of the chord will be congruent by CPCTC. The arcs will be congruent because their central angles are congruent.

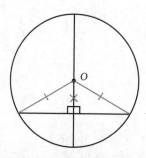

The next group usually expresses the relationship in terms of products, but those products come from cross-multiplying proportions because the relationships are based on similar triangles.

▶ When two chords intersect in a circle, the product of the lengths of the segments of one chord is equal to the products of the lengths of the segments of the other.

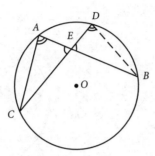

$\triangle EAC \sim \triangle EDB$, so $\dfrac{AE}{DE} = \dfrac{EC}{EB}$ and $AE \cdot EB = DE \cdot EC$

▶ If two secant segments are drawn to a circle from the same point, the product of the lengths of the external segment and the whole secant is the same for both secants.

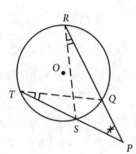

$\triangle PTQ \sim \triangle PRS$, so $\dfrac{PT}{PR} = \dfrac{PQ}{PS}$ and $PS \cdot PT = PQ \cdot PR$

▶ If a secant and a tangent are drawn to a circle from the same point, the product of the lengths of the external segment and the whole secant is equal to the length of the tangent segment squared.

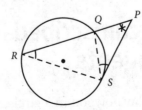

$\triangle PRS \sim \triangle PSQ$, so $\dfrac{PR}{PS} = \dfrac{PS}{PQ}$ and $PQ \cdot PR = PS^2$

## EXERCISE 12.4

**Apply the information in this section to find the lengths of the requested segments. Use the figure below for questions 1 through 5.**

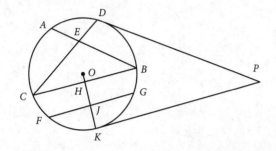

**Use the figure below for questions 6 through 10.**

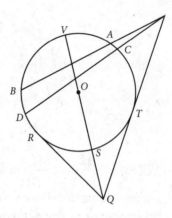

1. If $AE = 5$, $EB = 8$, and $DE = 4$, find the value of $EC$.

2. $\overline{OK} \perp \overline{FG}$. If $FG = 12$ and $OK = 10$, how long is $OJ$?

3. Use the information in question 2 to find the length of $JK$.

4. If $PK = 24$, and $\overline{OP}$ is drawn, find the length of $\overline{OP}$ to the nearest tenth.

5. If $\overline{DK}$ were drawn, what conclusion could you draw about $\angle PDK$ and $\angle PKD$?

6. If $PA = 9$, $AB = 16$, and $PD = 27$, how long is $\overline{PC}$?

7. If $PD = 27$ and $CD = 18\frac{2}{3}$, find $PT$.

8. If $QR = 12$ and $QV = 24$, find $QS$.

9. Using the information in question 8, find the diameter of circle $O$.

10. When $\overline{VS}$ intersects $\overline{AB}$, the segments of $AB$ can be represented by $x$ and $x + 1$, and their product is 56. When $\overline{VS}$ intersects $\overline{CD}$, the segments of $CD$ can be represented as $x + 1$ and $x + 2$, and their product is 72. Find the lengths of $AB$ and $CD$.

# Inscribed and Circumscribed Polygons and Constructing Tangents

When you constructed a regular polygon in a circle, you *inscribed* the polygon in the circle. You constructed it in such a way that the vertices were points on the circle. Each interior angle of the polygon was an inscribed angle of the circle. Each side of the polygon was a chord. The polygon is inside the circle, and as large as it can be without breaking out of the circle.

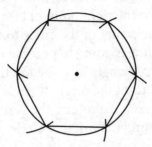

If, on the other hand, a polygon is *circumscribed* about a circle, each of its sides is tangent to the circle. The interior angles of the polygon are angles formed by two tangents. The circle is within the polygon, as large as it can be without breaking through the sides of the polygon.

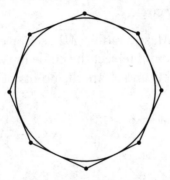

Inscribing a polygon in a circle usually involves constructions that divide the circle into arcs of equal measure and then connecting some or all of the points that define the arcs with chords. Circumscribing a polygon about a circle will require the construction of tangent lines. There are two constructions that produce tangent lines, depending on whether you are given the point of tangency or a point outside the circle.

The construction of a tangent to a point on the circle relies on the fact that a radius or diameter drawn to the point of tangency is perpendicular to the tangent. If you are given the point of tangency, you can construct a tangent by extending the radius through the point of tangency and then constructing a perpendicular to the radius at the given point.

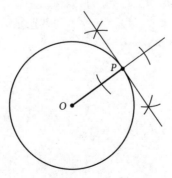

Constructing a tangent to the circle from a point outside the circle is not as simple, however, because you do not know where the point of tangency should be. This construction will still rely on the radii perpendicular to the tangent, but finding the right radius to the right point of tangency will take a little more work. Start by connecting the center of the circle to the point from which the tangent will be drawn. This line is going to become the diameter of a new circle and the hypotenuse of a right triangle inscribed in that circle. The legs of that right triangle will be the radius of the original circle and the tangent we want. To make that happen, we have to make sure the vertex of the right angle is a point on both the new circle and the original circle:

EXAMPLE

▶ Draw a line segment connecting the center of the circle, $O$, to the point outside, $P$.

▶ Construct the perpendicular bisector of $\overline{OP}$. Call the midpoint $M$. $M$ will be the center of the new circle.

▶ With compass point at $M$, and radius $\overline{MP}$, scribe a circle. Circle $M$ should pass through $P$ and $O$, and intersect the circle in two points. Those two points, which we'll label $Q$ and $R$, are the points of tangency.

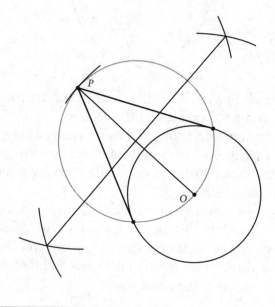

## EXERCISE 12.5

**Apply the skills of construction that you have learned to complete the following exercises.**

**1.** Construct a tangent to circle *C* at point *A*.

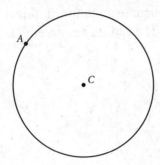

**2.** Construct two tangents from point *P* to circle *O*,

P •

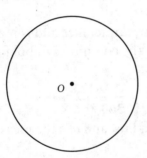

**3.** Inscribe a regular hexagon in circle *O* below.

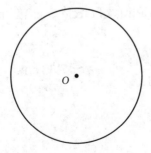

**4.** Construct two tangents from point *P* to circle *O* in the figure below.

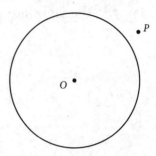

**5.** For each tangent constructed in question 4, use your compass to mark off a segment of the tangent line for which the point of tangency is a midpoint. Label these segments $\overline{PQ}$ and $\overline{PR}$.

**6.** Construct tangents to circle *O* from *Q* and from *R*.

**7.** Circumscribe a square about the circle.

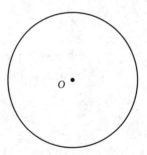

# Arc Measure, Arc Length, and Area of a Sector

Arc measure and arc length are related but should not be confused. The measure of an arc is a number of degrees equal to the measure of the central angle that intercepts that arc. Because it measures an amount of rotation, it is the same in small circles as in large circles. The radius has no effect.

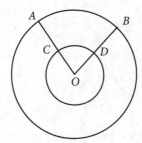

The measure of an arc is equal to the measure of the central angle that intercepts it. Arc measure is given in degrees.

But the size of the radius does make a difference in arc length. The length of an arc is a fraction of the circumference and depends on both the amount of rotation and the radius. An arc of a certain measure has a greater length in a large circle than in a small one.

Arc length is a fraction of the circumference and is measured in units of length like centimeters or inches. The common symbol for arc length is $s$.

$$s = \frac{a}{360°} \cdot 2\pi r$$

where $s$ is the arc length, $r$ is the radius, and $a$ is the measure of the arc or its central angle.

**EXAMPLE**

In circle $O$, arc $\overset{\frown}{AB}$ measures 30°. If the radius of circle $O$ is 14 cm, what is the length of arc $\overset{\frown}{AB}$?

$$s = \frac{a}{360°} \cdot 2\pi r = \frac{30°}{360°} \cdot 2\pi \cdot 14 = \frac{28\pi}{12} = \frac{7\pi}{3} \approx 7.33\,\text{cm}$$

Just as arc length is a fraction of the circumference of a circle, the area of a sector is a fraction of the area of the circle. Both depend upon the measure of the central angle or the intercepted arc.

$$A_{sector} = \frac{a}{360°} \cdot \pi r^2$$

where $a$ is the measure of the central angle or its arc and $r$ is the radius.

EXAMPLE

In circle $O$, arc $\overset{\frown}{AB}$ measures 30°. If the radius of circle $O$ is 14 cm, what is the area of the sector defined by $\angle AOB$?

$$A_{sector} = \frac{a}{360°} \cdot \pi r^2 = \frac{30°}{360°} \cdot \pi \cdot (14)^2 = \frac{196\pi}{12} = \frac{49\pi}{3} \approx 51.31 \text{ cm}^2$$

## EXERCISE 12.6

**Use the information in this section to answer these questions about arc length and area of sectors. Use the figure below for questions 1 through 3.**

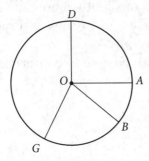

1. If m$\angle AOB = 40°$ and $OA = 12$ m, how long is $\overset{\frown}{AB}$? How long is $\overset{\frown}{ADB}$?

2. Find the area of the sector defined by $\angle DOG$ if m$\angle DOG = 120°$.

3. How much fencing would be required to enclose a garden in the shape of the sector defined by central angle $\angle DOG$?

4. Find the length of an arc that measures 57° if the radius of the circle is 30 cm.

5. Find the area of a sector defined by a 42° central angle in a circle with a diameter of 12 inches.

6. Which is longer: a 75° arc in a circle of radius 80 cm or an 80° arc in a circle of radius 72 cm?

7. Which has the larger area: a sector defined by a 30° central angle in a circle of radius 24 cm or a sector defined by a 60° central angle in a circle of radius 12 cm?

8. In the figure below, the two circles are concentric. The diameter of the inner circle is 6 cm and the diameter of the outer circle is 10 cm. If the central angle measures 40°, find the area of the shaded region.

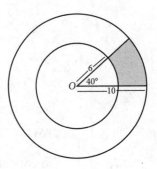

# Radian Measure

Angles are commonly measured in degrees. A *degree* is a unit of measure based on dividing a full rotation into 360 parts. Each part is 1 degree. Working in circles and carrying the idea of trig ratios into the algebraic world of functions leads to a different system of measurement of rotation and therefore of angles. Radian measure is built on a unit of one radian.

▶ **1 radian**   The measure of an angle whose intercepted arc has a length equal to the radius

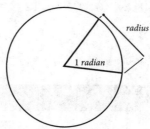

The definition of the radian is rooted in the circle. It uses the radius as a way to divide the rotation. The circumference of a circle is $C = 2\pi r$, so if it is divided into arcs equal to the radius, there will be $2\pi$, or slightly more than 6 arcs in the circumference. Each arc is the intercepted arc of an angle of 1 radian, which means that there are $2\pi$ radians in a full rotation.

The ratio of 1 unit to the whole rotation can be used to convert between radians and degrees.

$$\frac{\text{radians}}{2\pi} = \frac{\text{degrees}}{360}$$

An angle that measures 120° can be converted to radians by replacing degrees with 120 and solving the proportion.

$$\frac{\text{radians}}{2\pi} = \frac{120}{360}$$
$$360 \cdot \text{radians} = 120 \cdot 2\pi$$
$$\text{radians} = \frac{120 \cdot 2\pi}{360} = \frac{2\pi}{3}$$

Leave the result in terms of $\pi$ for an exact answer or use a calculator to approximate $\dfrac{2\pi}{3} \approx 2.09$ radians.

To convert radians to degrees, place the number of radians in the appropriate spot and solve the proportion.

$$\frac{\pi/4}{2\pi} = \frac{\text{degrees}}{360}$$
$$2\pi \cdot \text{degrees} = \frac{\pi}{4} \cdot 360$$
$$2\pi \cdot \text{degrees} = 90\pi$$
$$\text{degrees} = \frac{90\pi}{2\pi} = 45°$$

Using radian measure makes it easier to move from trigonometry in triangles to trigonometric functions, but it also makes arc length and area of sectors easier. Remember that both of those involve a fraction which was the number of degrees in the arc or central angle over 360 in the full rotation. If we use the symbol $\theta$ for the radian measure of the central angle and $2\pi$ for the full rotation, those formulas become simpler.

$$\text{Arc length } s = \frac{\theta}{2\pi} \cdot 2\pi r = \theta r$$

$$\text{Area of sector } A_{\text{sector}} = \frac{\theta}{2\pi} \cdot \pi r^2 = \frac{\theta r^2}{2}$$

## EXERCISE 12.7

**Use the proportional relationship between radian measure and degree measure to complete these exercises. In questions 1 through 3, convert the given degree measure to radians. Give an exact answer if possible, or round to the nearest tenth of a radian.**

**1.** 30°

**2.** 135°

**3.** 80°

**In questions 4 through 7, convert the given radian measure to degrees. Round to the nearest degree if necessary.**

**4.** $\frac{\pi}{3}$ radians

**5.** $\pi$ radians

**6.** 2.5 radians

**7.** 0.6 radians

**8.** Find the length of the arc intercepted by a central angle of $\frac{5\pi}{6}$ radians if the radius of the circle is 24 inches.

**9.** Find the number of radians in a central angle that intercepts an arc 108 cm long in a circle with diameter 144 cm.

**10.** Find the area of a sector defined by a central angle of $\frac{\pi}{2}$ radians if the diameter of the circle is 10 meters.

## CHAPTER 13

# Conic Sections

The circle is one member of a family of objects called **conic sections**. Each member of the family can be seen in the cross section of a cone, depending upon the angle of the cut. More importantly, for each conic section there is an equation which produces that conic when graphed. The equations also have similarities, so it makes sense to look at them as a group.

## Circles

When a cone is sliced parallel to the base of the cone, the cross section is a circle. The radius of the circle is larger when the slice is closer to the base and smaller when the slice is closer to the tip.

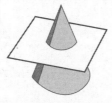

Just as you can find the equation that describes a graph that is a line, it is possible to find an equation that describes the graph of a circle. The equation comes from the definition of a circle: the set of all points at a fixed distance, called the *radius*, from a fixed point, called the *center*.

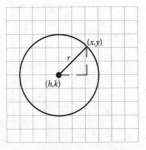

If the center is the point $(h, k)$ and the radius is $r$, then every point $(x, y)$ that sits on the circle is $r$ units from $(h, k)$. Apply the distance formula.

$$d = \sqrt{(x_2 - x_1)^2 + (y_2 - y_1)^2}$$
$$r = \sqrt{(x - h)^2 + (y - k)^2}$$

Square both sides to eliminate the radical and you have the equation of a circle.

$$(x - h)^2 + (y - k)^2 = r^2$$

**EXAMPLE**

▶ The equation of a circle of radius 3, centered at (4, 5) is $(x - 4)^2 + (y - 5)^2 = 9$.

If the center of the circle is at the origin, the equation simplifies to $x^2 + y^2 = r^2$.

**EXAMPLE**

▶ A circle of radius 5 centered at the origin has the equation $x^2 + y^2 = 25$.

**EXAMPLE**

▶ To sketch the graph of a circle with equation $(x + 2)^2 + (y - 3)^2 = 16$:

**Locate the center (−2, 3).**

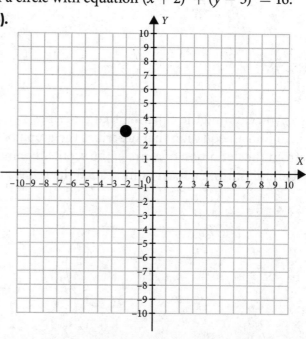

**From the center, count the radius of 4 up, down, left, and right, and mark each spot with a small arc.**

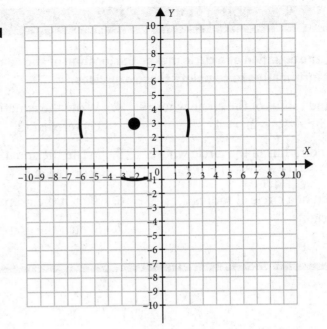

↓

**Connect the arcs with a circle.**

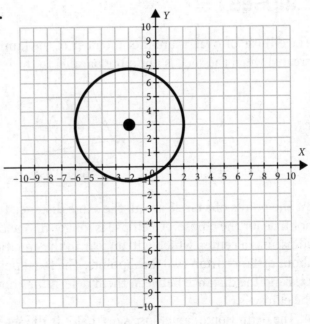

## EXERCISE 13.1

**In the proper format, the equation of a circle makes the location of the center and the length of the radius clear. Use that information to complete these exercises.**

1. Give the center and radius of the circle whose equation is $(x - 3)^2 + (y - 2)^2 = 4$.

2. Give the center and radius of the circle whose equation is $(x + 3)^2 + (y + 5)^2 = 121$.

3. Write the equation of a circle of radius 7 centered at the origin.

4. Write the equation of a circle of radius 10 centered at (2, 9).

5. Write the equation of a circle of radius 8 centered at (−7, −4).

6. Sketch the graph of $x^2 + (y - 3)^2 = 36$

7. Sketch the graph of $(x + 1)^2 + (y - 3)^2 = 64$

# Ellipses

The **ellipse**, or oval, is another member of the family of conic sections, but it is created when the slice is not quite parallel to the base but on a slight tilt.

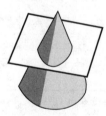

The points of a circle are all the same distance from the center, but that is not true for the ellipse. An ellipse is longer in one direction than the other. The ellipse has a center, which will appear in the equation, and the definition talks about distance from two focal points. The focal points always sit on the major axis and the distance between them determines how close to or far from a circle the ellipse is.

The definition of an ellipse says that it is the set of all points for which the total distance from the first focal point and from the second focal point is equal to a fixed constant.

> The closer the focal points are, the more like a circle the ellipse will look.

The equation of an ellipse is similar to the equation of a circle, but the ellipse does not have a single radius. It has a longer major axis and a shorter minor axis. Suppose you rewrite the equation of a circle by dividing each term by $r^2$. It would look like this:

$$\frac{(x - h)^2}{r^2} + \frac{(y - k)^2}{r^2} = 1.$$

The $r^2$ in the first denominator gives you information to help you sketch the graph. Move $r$ units left and $r$ units right from the center. The $r^2$ in the second denominator says move $r$ units up and $r$ units down. That's great for a circle, but for an ellipse, left-right movement will be different from up-down movement. Sometimes left-right will be larger; sometimes up-down will be larger. So the equation of the circle gets modified to produce the equation of an ellipse.

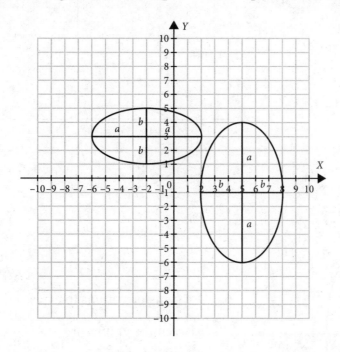

▶ Wider than it is tall: $\dfrac{(x-h)^2}{a^2} + \dfrac{(y-k)^2}{b^2} = 1$     $\dfrac{(x+2)^2}{4^2} + \dfrac{(y-3)^2}{2^2} = 1$

▶ Taller than it is wide: $\dfrac{(x-h)^2}{b^2} + \dfrac{(y-k)^2}{a^2} = 1$     $\dfrac{(x-5)^2}{3^2} + \dfrac{(y+1)^2}{5^2} = 1$

The letter $a$ denotes half of the major axis and the letter $b$ stands for half the minor axis, so the length of the major axis is $2a$ and the length of the minor axis is $2b$. The number in the denominator under $x$ tells you how far to move left and right, and the denominator under $y$ tells you how to move up and down.

To graph an ellipse, you can follow the same basic steps as graphing a circle, but the left-right movement will be different from the up-down movement.

## EXERCISE 13.2

**Use the information about ellipses presented in this section to complete the following questions.**

1. How long is the major axis

   of $\dfrac{(x-3)^2}{16} + \dfrac{(y+2)^2}{25} = 1$?

2. How long is the minor axis

   of $\dfrac{(x+1)^2}{81} + \dfrac{(y-7)^2}{49} = 1$?

3. Write the equation of the ellipse shown below.

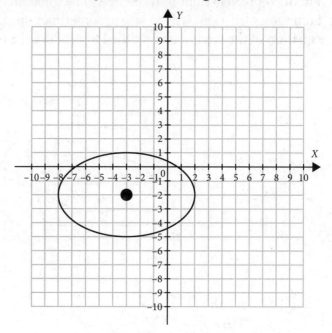

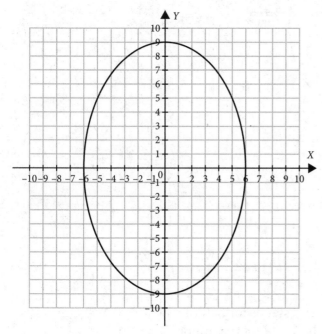

4. Write the equation of the ellipse shown below.

5. Use transformations to explain the difference

   between the graphs of $\dfrac{x^2}{16} + \dfrac{y^2}{9} = 1$ and

   $\dfrac{x^2}{9} + \dfrac{y^2}{16} = 1$.

6. Sketch the graph of $\dfrac{x^2}{36} + \dfrac{y^2}{4} = 1$.

7. Sketch the graph of $\dfrac{(x-2)^2}{9} + \dfrac{(y+3)^2}{25} = 1$.

8. Graph on the same axes: $x^2 + y^2 = 9$,

   $x^2 + y^2 = 25$, and $\dfrac{x^2}{9} + \dfrac{y^2}{25} = 1$.

9. If the area of an ellipse can be found using the formula $A = \pi ab$, by how much does the area of the ellipse in question 8 exceed the area of the smaller circle? By how much does the area of the larger circle exceed the area of the ellipse?

# Parabolas

The conic section called a **parabola** is created when the cone is sliced at a steeper angle so that rather than entering on one side and exiting on the other, the plane slicing through enters on one side and exits at the base. It's not vertical; that's saved for the fourth conic.

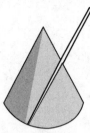

The parabola is defined as the set of all points equidistant from a point called the focus and a line called the **directrix**. The **vertex**, or turning point, is midway between the focus and the directrix.

You may already be familiar with the parabola as the shape of the graph of a quadratic equation $y = ax^2 + bx + c$. You can also look at what's called the **vertex form of the equation**, which is easy to graph and in which you may see some similarity to the other conics. It describes a parabola whose vertex is the point $(h, k)$. The sign of $a$ shows the direction in which the parabola opens and the absolute value of $a$ indicates the width of the parabola.

Parabola that opens up or opens down: $y - k = a(x - h)^2$

EXAMPLE

On the left of the figure below is the graph of $y - 5 = -2(x + 4)^2$ and on the right $y + 4 = (x - 3)^2$

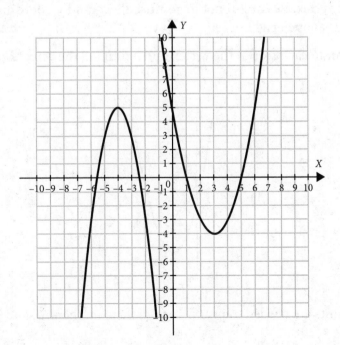

There is another possibility you may not have seen in algebra:

▶ **Parabola that opens right or left**    $x - h = a(y - k)^2$.

**EXAMPLE**

▶ On the upper left of the figure below is the graph of $x + 4 = -0.3(y - 6)^2$. On the lower right is $x + 3 = (y + 1)^2$.

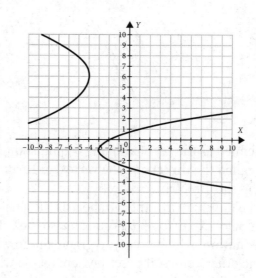

The larger the absolute value of $a$, the narrower the parabola will be. The actual value of $a$ depends on the distance between the focus and the vertex. The bigger that distance is, the smaller $a$ will be and the wider the parabola will be.

> If $a$ is positive, the parabola opens up or to the right. If $a$ is negative, the parabola opens down or to the left.

**EXAMPLE**

▶ Locate the vertex. Be careful not to confuse the $x$- and $y$-coordinates on parabolas that open right or left.

▶ Move 1 unit to each side of the vertex and $a$ units in the direction of opening.

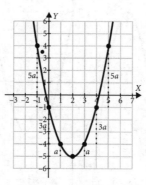

▶ If more points are needed, repeat using $3a$, then $5a$, then $7a$, and so on.

## EXERCISE 13.3

1. For each of the equations below, tell whether the parabola opens up, down, left, or right.

   a) $y - 6 = -3(x + 2)^2$
   b) $x + 4 = \frac{1}{2}(y - 3)^2$
   c) $y - 4 = 3(x + 5)^2$

2. Identify the vertex of each parabola.

   a) $y - 6 = -3(x + 2)^2$
   b) $x + 4 = \frac{1}{2}(y - 3)^2$
   c) $y - 4 = 3(x + 5)^2$

3. Write the equation of the parabola shown below.

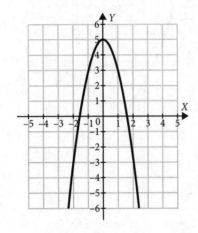

4. Write the equation of the parabola shown below.

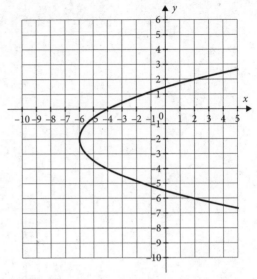

5. Sketch the graph of $y - 1 = (x + 3)^2$

6. Sketch the graph of $x - 3 = -2(y + 2)^2$

7. Sketch the graph of $y + 4 = 3(x - 1)^2$

# Hyperbolas

The final conic is the most dramatic shape. The **hyperbola** is created by slicing vertically through a double cone, creating two curves that open away from one another like wings.

The hyperbola has an equation very similar to the ellipse, but rather than adding the terms, it subtracts. There are a lot of variations on the equation, because of the length of axes and the direction of opening.

▶ **Opening left and right** $\dfrac{(x-h)^2}{a^2} - \dfrac{(y-k)^2}{b^2} = 1$

or $\dfrac{(x-h)^2}{b^2} - \dfrac{(y-k)^2}{a^2} = 1$

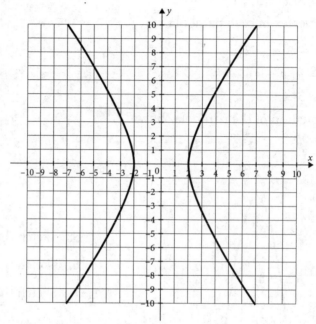

▶ **Opening up and down** $\dfrac{(y-k)^2}{a^2} - \dfrac{(x-h)^2}{b^2} = 1$

or $\dfrac{(y-k)^2}{b^2} - \dfrac{(x-h)^2}{a^2} = 1$

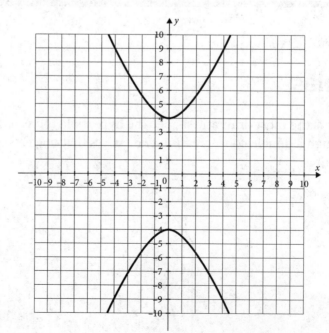

To sketch a hyperbola:

Locate the center.

↓

Count to left and right and mark the spot.

↓

Count up and down and mark the spot.

↓

Lightly sketch a rectangle whose sides pass through the marked points.

↓

Draw the diagonals of the rectangle and extend them.

↓                                                    ↓

If the equation is $x^2 - y^2$,
sketch the wings between the
diagonals, just touching the left
and right sides of the rectangle.

If the equation is $y^2 - x^2$,
sketch them touching the top
and bottom of the rectangle.

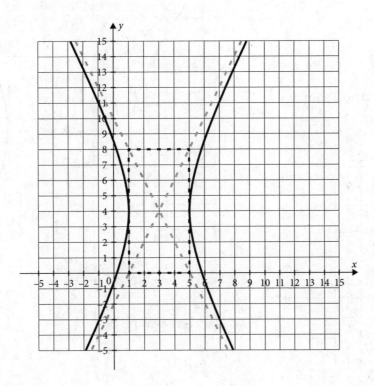

## EXERCISE 13-4

1. For each of the equations below, identify the center of the hyperbola.

   a) $\dfrac{x^2}{25} - \dfrac{(y-3)^2}{9} = 1$

   b) $\dfrac{(y-3)^2}{49} - \dfrac{(x+2)^2}{9} = 1$

   c) $\dfrac{(x-1)^2}{9} - \dfrac{(y+1)^2}{4} = 1$

2. Tell whether each hyperbola opens to the left and right or opens up and down.

   a) $\dfrac{x^2}{25} - \dfrac{(y-3)^2}{9} = 1$

   b) $\dfrac{(y-3)^2}{49} - \dfrac{(x+2)^2}{9} = 1$

   c) $\dfrac{(x-1)^2}{9} - \dfrac{(y+1)^2}{4} = 1$

3. Write the equation of the hyperbola shown below.

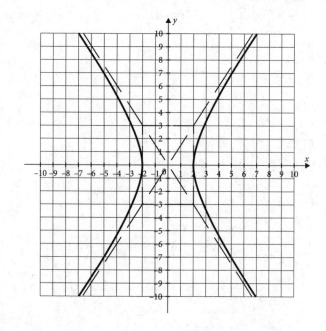

4. Write the equation of the hyperbola shown below.

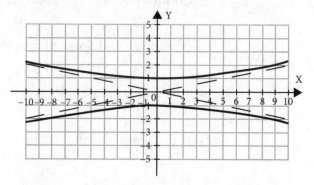

5. Write the equation of the hyperbola shown below.

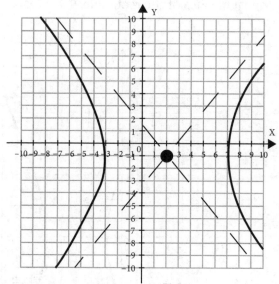

6. Sketch the graph of $\dfrac{x^2}{4} - \dfrac{y^2}{9} = 1$.

7. Sketch the graph of $\dfrac{y^2}{36} - \dfrac{x^2}{81} = 1$.

# Circles and Conics

These review questions bring together the ideas from Chapters 12 and 13. This is your chance to test your knowledge of circles and related conics. Answer all the questions, and try to express your thinking as clearly as you can. Use the figure below for questions 1 through 3.

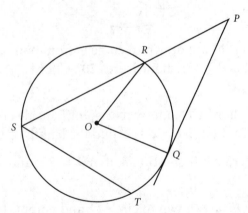

1. Name a chord.

2. Name an inscribed angle.

3. How is the angle is question 2 measured?

4. If the circumference of circle $O$ is $24\pi$, what is the radius of circle $O$?

5. The two circles in the figure below are concentric. The diameter of the smaller circle is half the diameter of the larger circle. What fraction of the large circle is shaded?

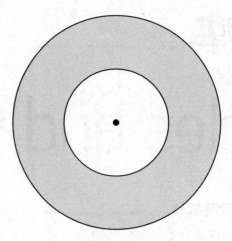

6. In the figure below, circle $C$ is internally tangent to circle $O$ and $O$ is a point on circle $C$. m$\angle OAE = 30°$. If $\overset{\frown}{OE}$ is $\pi$ cm long, how long is $\overset{\frown}{DC}$?

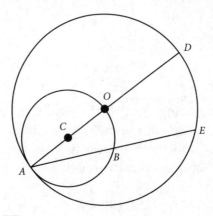

7. Tangents $\overline{PQ}$ and $\overline{PR}$ are drawn to circle $O$ from point $P$ outside the circle, forming $\angle RPQ$, which measures 20°. Chord $\overline{QR}$ is drawn. Find the measure of $\angle PQR$.

8. Chord $\overline{AB}$ and chord $\overline{CD}$ intersect at point $E$ inside circle $O$. If $AE = x$, $EB = 3x$, $CE = 2x$, and $ED = x + 2$, find the length of $\overline{CD}$.

9. In the circle described in question 8, if $\overset{\frown}{AD} = 72°$ and $\overset{\frown}{BC} = 64°$, find m$\angle AEC$

10. Radii $\overline{OA}$ and $\overline{OB}$ are drawn in circle $O$, and point $C$ lies on circle $O$. $\overset{\frown}{ACB}$ is a major arc. Acute angle $\angle AOB$ is a central angle that measure 48°. Find the measure of $\angle ACB$.

11. Secants $\overleftrightarrow{PA}$ and $\overleftrightarrow{PD}$ are drawn to circle $O$ from point $P$ outside the circle, as shown in the figure below. $PB = 3$ cm, $PC = 5$ cm, and $AB = 17$ cm. Find the length of $\overline{CD}$.

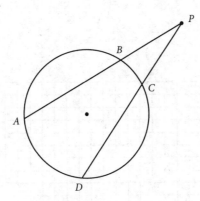

12. Construct tangents to circle *O* from point *P* using compass and straightedge.

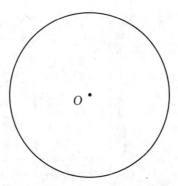

13. If regular hexagon *ABCDEF* is inscribed in circle *O*, find the measure of $\overset{\frown}{BDF}$.

14. From a point *P* outside circle *O* described in question 13, two tangents are drawn, one to *B* and one to *F*. Find the measure of ∠*BPF*.

15. In a certain circle *C*, an angle of 85°, formed by radii $\overline{CA}$ and $\overline{CB}$, defines a sector with an area of 34 square inches. What is the length of arc $\overset{\frown}{AB}$?

16. What is the radian equivalent of 120°?

17. What is the equivalent degree measure for an angle of $\dfrac{3\pi}{4}$ radians?

**18.** Sketch a graph of $x^2 + y^2 = 36$.

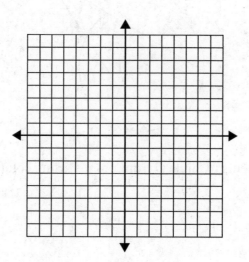

**19.** Write the equation of the circle shown below.

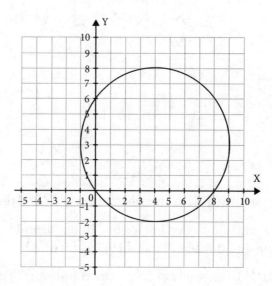

**20.** Identify each equation below as circle, ellipse, parabola, or hyperbola.

a) $\dfrac{x^2}{81} - \dfrac{y^2}{49} = 1$

b) $(x + 4)^2 + (y - 7)^2 = 4$

c) $\dfrac{(x - 1)^2}{16} + \dfrac{(y - 2)^2}{9} = 1$

d) $x - 3 = 4(y - 1)^2$

## CHAPTER 14

# Three-Dimensional Figures

Much of the geometry studied in school is plane geometry, focused on points, lines, and two-dimensional figures like polygons and circles. The world we live in is not two-dimensional, of course, and so it makes sense to spend some time looking at three-dimensional objects. Commonly referred to by the generic term *solids*, they are not actually filled, but these three-dimensional objects can be used for modeling real situations.

## Polyhedrons

A **polyhedron** is a three-dimensional object formed by joining polygons at their edges. A polyhedron has **faces**, which are polygons, or sections of a plane. It has **edges**, which are the line segments where two faces meet, and **vertices**, or points where several faces come together. **Convex polyhedrons**, like convex polygons, have no dents or holes. Polyhedrons are named by the number of faces, similar to the way polygons are named by the number of sides. A polygon with five sides is a pentagon; a polyhedron with five faces is a pentahedron. This general naming system is often replaced by more common names. A regular hexahedron is more likely to be called a cube.

In the 18th century, Swiss mathematician Leonhard Euler developed a formula to relate the number of faces, number of edges, and number of vertices in any convex polyhedron. With a bit of observation, you may see it too. Think about a cube.

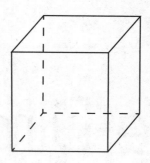

How many faces? How many edges? How many vertices?
Now look at a square pyramid.

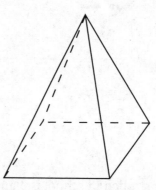

How many faces? How many edges? How many vertices?
Can you see it yet? Look at other polyhedrons. Count faces, edges, and vertices. Add the number of faces to the number of vertices. Notice anything?

**Euler's Polyhedron Formula**     Faces + Vertices = Edges + 2

You'll see the formula rearranged in different ways, like Faces + Vertices – Edges = 2, but choose the one that's easiest for you to remember.

There are other three-dimensional figures that do not meet the definition of polyhedron, usually because one or more of their surfaces are not polygons. Think of a sphere or a cone or a cylinder. These shapes that curve don't qualify as polyhedrons, but many ideas used in dealing with polyhedrons can be applied, with a little modification, to some of their curved cousins.

## EXERCISE 14.1

These exercises focus on the names of polyhedrons and on Euler's formula. Use that information to answer each question. Sketch the polyhedron if you find it helpful. Formal names for polyhedrons use prefixes that indicate the number of faces: tetra- for 4, penta- for 5, hexa- for 6, and so on. Many polyhedrons have simpler common names.

**1.** What is the common name for a tetrahedron?

**2.** What is the formal name for a square pyramid?

**3.** What is the formal name for a pentagonal prism?

**4.** If a polyhedron has 8 faces and 12 vertices, how many edges will it have?

**5.** If a polyhedron has 6 edges and 4 vertices, how many faces will it have?

**6.** If a polyhedron has 6 faces and 12 edges, how many vertices will it have?

**7.** How many faces, vertices, and edges does a hexagonal prism have?

# Surface Areas of Prisms and Cylinders

**Surface area** is just the total of the area of all the faces of a polyhedron. If you're looking at a cube, you see 6 faces, each of them a square with the same side length. You find the area of one of the squares, by squaring the length of its side, and multiply by 6.

As the polyhedrons become more complicated, of course, it's not always that easy. How many faces? What shape? Are they all the same shape? Do you have the dimensions you need to plunge right into finding areas? That's why many people like to take the time to decompose a polyhedron into a flat image, called a **net**, which can fold up into the polyhedron. A net allows you to see each of the polygons that form the polyhedron and calculate area piece by piece.

A net for a cube might look like this:

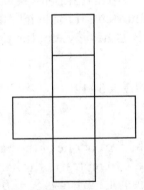

A net for a square pyramid might look like this:

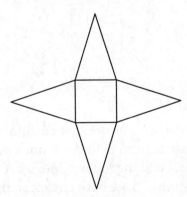

A **prism** is a polyhedron with two identical, parallel bases, surrounded by lateral sides that are parallelograms. If the bases are directly above one another and the lateral sides are rectangles, the prism is a **right prism**. The prism takes its name from the shape of the bases. If the bases are pentagons, the prism is a **pentagonal prism**. A **triangular prism** has bases that are triangles.

If the bases of a prism are regular polygons, the lateral area can be found by finding the area of one parallelogram and multiplying by the number of sides, but if the bases are not regular, the area of each parallelogram may need to be calculated separately. In that case, a well-labeled net is helpful.

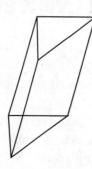 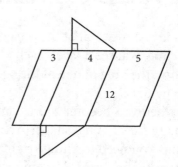

▸ The triangular prism above decomposes into a net with 3 rectangles and 2 congruent right triangles. The triangles are 3-4-5 right triangles, and the rectangles are 3 by 12, 4 by 12, and 5 by 12. The total surface area is $2\left(\frac{1}{2} \cdot 3 \cdot 4\right) + 3 \cdot 12 + 4 \cdot 12 + 5 \cdot 12$. Before evaluating that expression, stop to notice a possible shortcut. If the height of 12 is factored out of the last three terms, the result is the height of the prism times the perimeter of the base.

$$2\left(\tfrac{1}{2} \cdot 3 \cdot 4\right) + 3 \cdot 12 + 4 \cdot 12 + 5 \cdot 12$$
$$2\left(\tfrac{1}{2} \cdot 3 \cdot 4\right) + 12(3 + 4 + 5)$$

▸ If we use $B$ to represent the area of one of the bases, $P$ to represent the perimeter of the base, and $h$ to represent the height of the prism, the formula for the surface area of a prism is $S = 2B + Ph$.

$$2\left(\tfrac{1}{2} \cdot 3 \cdot 4\right) + 12(3 + 4 + 5) = \cancel{2}\left(\tfrac{1}{\cancel{2}} \cdot 3 \cdot 4\right) + 12(12)$$
$$= 12 + 144$$
$$= 156$$

**Cylinders** are all around, as cans or other containers, or the shape of everything from poles to pencils. A cylinder is not a polyhedron because of its curved lateral surface, but breaking it down into a net shows how the surface area can still be found. The bases are both circles, so the area of the bases is $2\pi r^2$. The lateral area can be imagined as unrolling the label on a cylindrical can. It unwraps to a rectangle, with one dimension equal to the height of the cylinder and the other dimension equal to the circumference of the circle. The surface area of a cylinder is $S = 2\pi r^2 + 2\pi rh$.

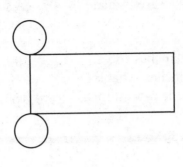

EXAMPLE If a cylinder has a surface area of $4,800\pi$ square cm and a height of 50 cm, then the equation for surface area $S = 2\pi r^2 + 2\pi rh$ becomes $4,800\pi = 2\pi r^2 + 2\pi r(50)$, which simplifies to $0 = 2\pi r^2 + 100\pi r - 4,800\pi$ or $0 = r^2 + 50r - 2,400$. Factor to get $(r + 80)(r - 30) = 0$, which produces two solutions. The negative solution, $r = -80$, is rejected, so the radius is 30 cm.

## EXERCISE 14.2

**Nets can be helpful when you are trying to find surface area, so draw a net whenever you think it might be helpful as you work these questions. Round as directed and be certain to include appropriate units.**

**1.** Which of these is a net for a triangular pyramid?

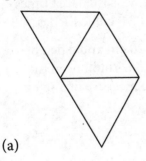

(a)

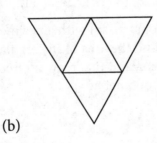

(b)

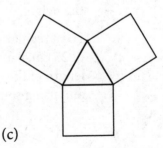

(c)

**2.** Draw a net for a hexagonal prism.

**3.** Find the surface area of a pentagonal prism if the bases are regular pentagons with perimeter = 20 cm and apothem = 2.75 cm and the height of the prism is 25 cm.

**4.** Find the surface area of a cylinder with a radius of 12 inches and a height of 18 inches.

**5.** If a hexagonal prism has a surface area of 1,100.55 cm² and a perimeter of 48 cm, and each base has an area of 166.28 cm², what is the height of the prism, to the nearest square centimeter?

**6.** If a cylinder has a height of 9 inches and a surface area of $104\pi$ square inches, what is the radius of the cylinder to the nearest inch?

**7.** A square prism and a cylinder have the same height. The distance from the center of the square base to its vertex is equal to the radius of the circular base of the cylinder. Which object has the greater surface area? Explain your reasoning.

# Volumes of Prisms and Cylinders

Imagine that you've been giving the task of packing children's alphabet blocks into a box. You could just dump them in and hope for the best, but the blocks are cubes, 1 inch on each edge, so they will pack in nicely. If you have the patience to pack them, the blocks will settle into a layer, and the number of blocks in the layer will depend on the length and width of the box. In a box that measures 12 inches long, 8 inches wide, and 10 inches high, the bottom layer will have 8 rows of 12 blocks, or 96 blocks per layer, and you'll be able to make 10 such layers, for a total of 960 blocks. The volume of the box is 960 cubic inches.

For a rectangular prism, the area of the base is length times width, so $V = lwh$, but for other prisms, finding volume will not always be as easy as length times width times height. The calculation of the base area will be more complicated, but volume is the area of the base times the height. The volume of a prism is $V = Bh$.

EXAMPLE

To find the volume of a rectangular prism with bases that measure 8 inches wide and 14 inches long, and lateral faces that have a height of 27 inches, the area of the base is $8 \cdot 14$, and the volume is that area times the height. $V = Bh = (8 \cdot 14) \cdot 27 = 112 \cdot 27 = 3{,}024$ cubic inches.

For a regular octagonal prism with a perimeter of 120 cm and a height of 25 cm, pause to find the area of the regular octagon. You'll need the length of the *apothem*, and that requires finding the measure of angles and using some trigonometry.

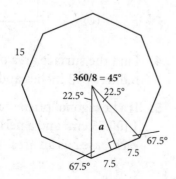

$$\tan(67.5°) = \frac{a}{7.5}$$
$$a = 7.5\tan(67.5°)$$
$$a \approx 18.1$$

▶ Once you have the apothem, the area of the octagon is $\frac{1}{2}aP = \frac{1}{2}(18.1)(120) = 1{,}086$. Finally, you can calculate $V = Bh = (1{,}086)(25) = 27{,}150$ cubic centimeters.

The idea of volume as area of the base times the height can be applied to cylinders with a small adjustment. The base of a cylinder is a circle, so the area of the base is $\pi r^2$, and the volume of the cylinder is $V = \pi r^2 h$.

**EXAMPLE**

▶ If a cylinder has a volume of $540\pi$ cubic centimeters and a height of 15 cm, the radius can be found by substituting known values into the formula and solving.

$$V = \pi r^2 h$$
$$540\pi = \pi r^2 \cdot 15$$
$$\frac{540\pi}{15\pi} = r^2$$
$$36 = r^2$$
$$r = 6$$

## EXERCISE 14.3

**These questions concern the volumes of prisms and cylinders. Round answers as directed, and be sure to include appropriate units.**

1.  Find the volume, to the nearest tenth of a cubic inch, of a prism whose base is a regular hexagon 8 inches on a side, if the height of the prism is 15 inches.

2.  Find the volume of a cylinder with a diameter of 20 cm and a height of 12 cm.

3.  If a regular pentagonal prism has a volume of 25,807.2 cubic feet and a base area of 1,075.3 square feet, find the height to the nearest foot.

4.  If a cylinder has a radius of 81 cm and a volume of 787,320 cm³, find the height to the nearest centimeter.

**5.** If a cylinder has a surface area of $1056\pi$ square inches and a diameter of 24 inches, find the volume in terms of $\pi$.

**6.** Two triangular prisms have right triangle bases. One has a base that is a 3-4-5 right triangle, and the other has a base that is a 5-12-13 right triangle. If both prisms have the same volume, what is the ratio of the height of the first prism to the height of the second?

**7.** If the radius of a cylinder is doubled, what change should be made to the height if you want the new cylinder to have the same volume as the original?

# Surface Areas of Pyramids and Cones

A **pyramid** is a polyhedron with one polygonal base and lateral sides that are triangles which tip in so that their vertices all meet at a point.

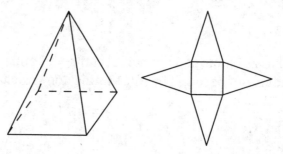

The net for a pyramid shows the base surrounded by triangles. The surface area of the pyramid is equal to the area of the base, $B$, plus the total of the areas of all the triangles. The area of each triangle is $\frac{1}{2}bh$, but look carefully at the pyramid before you start to calculate. The base of each triangle is an edge of the base polygon, but the height of the triangle is not the height of the pyramid. The height of the pyramid is measured from the top point of the pyramid perpendicular to the base. The height of the triangle, which is called the *slant height* and designated as $l$, is the hypotenuse of a right triangle formed by the apothem of the base, the height of the pyramid, and the slant height. The *apothem* is the measure from the center of the base, perpendicular to an edge.

The slant height is $l = \sqrt{a^2 + h^2}$.

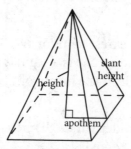

If $B$ is the area of the base, $l$ is the slant height, and $P$ is the perimeter of the base, the surface area of a pyramid is $S = B + \frac{1}{2}lP$.

EXAMPLE

To find the surface area of a regular hexagonal pyramid with a base perimeter of 60 cm and a height of 11 cm, first find the apothem of the hexagonal base.

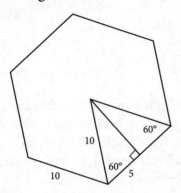

$$a^2 + 5^2 = 10^2$$
$$a^2 + 25 = 100$$
$$a^2 = 75$$
$$a = 5\sqrt{3}$$

▶ With the apothem and the perimeter, it's possible to calculate the area of the base.

$$B = \tfrac{1}{2}aP = \tfrac{1}{2}\left(5\sqrt{3}\right)(60) = 150\sqrt{3}$$

▶ The base is surrounded by 6 identical triangles, each with a base of 10 cm, but we must calculate the slant height.

$$l = \sqrt{\left(5\sqrt{3}\right)^2 + 11^2} = \sqrt{75 + 121} = \sqrt{196} = 14\,\text{cm}$$

▶ Finally, the surface area is $S = B + \tfrac{1}{2}lP = 150\sqrt{3} + \tfrac{1}{2}(14)(60) =$

$150\sqrt{3} + 420 \approx 679.8$ cubic centimeters.

Applying similar thinking to a cone, you can see that the slant height is the hypotenuse of a right triangle formed by the radius of the base, the height of the cone, and the slant height: $l = \sqrt{r^2 + h^2}$.

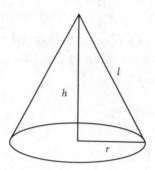

The area of the base of the cone is easy: $A = \pi r^2$. Finding the lateral area of the cone is not as simple. If you unroll it for a net, you see that it is part, but not all, of a circle. But what part? And what is the radius of that circle? If you look at the process of unrolling the lateral area, you'll see that the radius of this partial circle is actually the slant height, $l$. Realize too that when you roll this partial circle into a cone, it fits right around the circumference of the base. That means that the arc length of this partial circle is the circumference of the base, so $s = 2\pi r$.

If it were a full circle, its circumference would be $C = 2\pi l$. The portion of the circle that forms the lateral area has an arc length that is $\dfrac{2\pi r}{2\pi l} = \dfrac{r}{l}$ of the full circumference, so it will have $\dfrac{r}{l}$ of the full area: $\dfrac{r}{l} \cdot \pi l^2 = \pi r l$. To get the total surface area, add the base area and the lateral area: $S = \pi r^2 + \pi r l$. The surface area of a cone is $S = \pi r^2 + \pi r l$.

A cone with a radius of 6 inches and height of 8 inches has a slant height of $l = \sqrt{r^2 + h^2} = \sqrt{6^2 + 8^2} = \sqrt{36 + 64} = \sqrt{100} = 10$ inches. Its surface area is $S = \pi r^2 + \pi r l = \pi \cdot 6^2 + \pi \cdot 6 \cdot 10 = 36\pi + 60\pi = 96\pi$ square inches.

## EXERCISE 14.4

**Surface areas of pyramids and cones may require you to find the slant height. Don't confuse slant height with the height. Work each question carefully, round as directed and be sure to include units.**

1. Find the surface area of a regular pentagonal pyramid with a base area of 247.75 square centimeters, a perimeter of 60 centimeters, and a slant height of 26.3 cm.

2. Find the surface area of a cone with a diameter of 18 inches and a slant height of 24 inches.

3. A pyramid has a regular hexagon as its base. The perimeter of the hexagon is 96 inches, and the height of the pyramid is 8 inches. Find the surface area of the pyramid.

4. If a pyramid has a surface area of 1,024.5 square inches, a base area of 377 square

inches, and a perimeter of 70 inches, find the slant height to the nearest tenth of an inch.

5. Which cone has the larger surface area: one with a radius of 3 cm and a height of 4 cm, or one with a radius of 4 cm and a height of 3 cm?

6. Two square pyramids have the same square as a base. One has a height of 1 foot, and the

other has a height of 2 feet. If each side of square base is 6 feet long, how much greater is the surface area of the taller pyramid?

7. Which has a larger surface area: a square pyramid with a perimeter of 40 cm and a height of 20 cm, or a cone with a radius of 10 cm and a height of 20 cm?

# Volumes of Pyramids and Cones

Finding volumes of pyramids and cones is very similar to finding volumes of prisms and cylinders, but it is necessary to account for the fact that pyramids and cones narrow to a point and therefore will have a smaller volume. Fortunately, the adjustment is always the same. The volume of the pyramid is one-third the volume of the prism with the same base area and height, and the volume of a cone is one-third the volume of the cylinder with the same base and height.

▶ **Volume of a pyramid** $$V = \frac{1}{3}Bh$$

EXAMPLE

▶ A square pyramid, 15 cm on each side, with a height of 20 cm has a volume of:

$V = \frac{1}{3}Bh = \frac{1}{3}(15^2)(20) = \frac{1}{3}(225)(20) = 1{,}500$ cubic centimeters.

▶ **Volume of a cone** $$V = \frac{1}{3}\pi r^2 h$$

EXAMPLE

▶ A cone with a radius of 6 inches and height of 8 inches has a volume of:

$V = \frac{1}{3}\pi r^2 h = \frac{1}{3}\pi(6)^2(8) = \frac{1}{3}\pi \cdot 36 \cdot 8 = 96\pi$ cubic inches.

## EXERCISE 14.5

These questions focus on the volumes of pyramids and cones. Apply the formulas of this section to solve each problem. Include appropriate units.

1.  Find the volume of a triangular pyramid 18 cm high whose base is a 3 cm-4 cm-5 cm right triangle.

2.  Find the volume of a cone with a radius of 8 inches and a height of 12 inches.

3.  Find the radius of a cone that has a height of 25 cm and a volume of $300\pi$ cubic centimeters.

4.  Find the height of a pyramid with a base area of 42 square inches and a volume of 154 cubic centimeters.

5.  Find the volume of a cone with a slant height of 5 cm if the cone has a surface area of $24\pi$ square centimeters.

6.  A square pyramid has a volume of 720 cubic centimeters and a height of 15 cm. Find the surface area of the pyramid.

7.  When a cone is placed inside a cylinder with the same base and height as the cone, there is $300\pi$ cubic inches of empty space. If the height is 18 inches, find the radius of the cone.

# Spheres

**Spheres** get a category of their own. Their ball shape has no bases and no points, no flat faces. Deriving the formulas for the surface area and volume of a sphere requires calculus, so we won't do that here, but there are some ways you can see the logic of them. Carefully peel an orange. Try to keep the peel in as few pieces as possible, and then flatten them on a surface. Break the orange into two halves. The cross section is a circle with area $\pi r^2$ for whatever the radius of the orange is. Place the two halves, circle side down, on the flattened peel and you should see they take up about half the space of the peel. You could fit two more of those circles. The surface area of a sphere with radius $r$ is $S = 4\pi r^2$.

EXAMPLE

> If a sphere has a radius of 5 inches, its surface area is
> $S = 4\pi r^2 = 4\pi \cdot 5^2 = 100\pi$ square inches.

   If you wanted to create a container in which to place your orange, it would have to be at least as wide as the diameter of your orange, and at least as tall as that diameter. If you create a cylinder with the diameter of the orange, or $2r$, and a height of $2r$, the volume of that cylinder is $V = \pi r^2 h = (\pi r^2)(2r) = 2\pi r^3$.

Your whole orange would fit inside, but there would be space left over. The cylindrical container has a volume a bit more, but not a lot more, than the volume of the orange. The volume of a sphere with radius $r$ is $V = \frac{4}{3}\pi r^3$.

EXAMPLE

A sphere with a radius of 3 inches will have a volume of
$V = \frac{4}{3}\pi r^3 = \frac{4}{3}\pi \cdot 3^3 = 36\pi$ cubic inches.

## EXERCISE 14.6

**Spheres are all around and yet finding a way to calculate their surface area and volume is a challenging task. Use the formulas of this section to answer each exercise. Be sure to show units.**

1. Find the surface area of a sphere of radius 14 cm.

2. Find the volume of a sphere of radius 4 inches.

3. What is the radius of a sphere with a volume of $288\pi$ cubic centimeters?

4. What is the diameter of a sphere whose surface area is $324\pi$ square inches?

5. If the volume of a sphere is $2,304\pi$ cubic meters, what is its surface area?

6. For what radius, in centimeters, is the surface area of a sphere, in square centimeters, equal to the volume of that sphere, in cubic centimeters?

7. A regulation soccer ball is not actually a sphere but is formed from 12 pentagonal faces and 20 hexagonal faces. It has a diameter of approximately 22 cm. If it were in fact spherical, what would its volume be?

# Modeling with Three-Dimensional Figures

Three-dimensional figures can be used to model real-life situations because real-life situations involve three-dimensional objects. Those objects may not be perfect cylinders or precisely a hexagonal pyramid, but they may be close enough to allow reasonable estimates.

One of the ways to estimate volumes of irregularly shaped objects is to think about cross sections. If you have one slice from a loaf of bread, you can measure its length, width, and thickness to find the volume of that slice. Then if the slices are all the same shape and size, you can find the volume of the loaf by multiplying the volume of one slice by the number of slices. If the slices are not all the same, you might divide the loaf into sections, with similar slices throughout each section.

Many common polyhedrons can be created by rotating common polygons. Rotate:

▶ A rectangle about one of its sides and you produce a cylinder.

▶ A right triangle about one of its legs and you produce a cone.

▶ A circle about one of its diameters and you produce a sphere.

Here too, you may want to break an object into parts. The body of a rocket might be modeled by placing a right triangle atop a rectangle and rotating the combination about a side.

## EXERCISE 14.7

These questions draw together many of the ideas of this chapter. Work each exercise carefully, drawing diagrams wherever helpful. Don't forget units.

1. Janet's ice cream shop offers a child-size cone with a single scoop of ice cream. Assume the scoop of ice cream is a sphere with a volume of $36\pi$ cubic centimeters. Find the diameter of the scoop.

2. Because Janet knows from experience that little children can't always eat all their ice cream before it melts, she wants to be sure the cone she uses will catch as much of the melted ice cream as possible. If the diameter of the cone is the same as the diameter of the scoop you found in question 1, what should the height of the cone be so that it can hold the entire $36\pi$ cubic centimeters of melted ice cream?

3. In New York City, buildings of more than six stories generally have a rooftop water tank and pumping system to provide adequate water pressure to building residents. The typical tank holds 10,000 gallons of water. If 7.48 gallons of water will fit in 1 cubic foot, find the necessary volume of the cylindrical tank. If the tank is 12 feet high, what will its diameter be?

4. A manufacturer of cylindrical cans seeks to minimize the amount of material used in producing a can of a given volume. If the corporation is manufacturing a cylindrical can with a volume of 250 cubic centimeters, and considering designs with radii of 3 cm or 4 cm, which would have the smaller surface area?

5. Planet Earth is not a perfect sphere, but it is close enough that we can use a sphere as a model of the Earth. If the radius of the Earth is 3,959 miles, and 71% of the Earth is covered by water, how many square miles are water covered?

6. The Great Pyramid of Giza, Egypt, and the Great Pyramid of Cholula, Mexico, are both square pyramids. The Pyramid of Giza is 146.5 meters high, with a base approximately 230 meters on a side. The Pyramid of Cholula is shorter, at 55 meters, but has a base 400 meters on a side. Which pyramid has the larger volume?

7. Two types of silos are commonly used for grain storage. One is a cylindrical structure topped with a hemispheric roof. The other is also cylindrical, but its roof is a cone. If a hemisphere silo and a conical silo are built with identical cylindrical sections, and also have the same volume, what should be the height of the cone?

# Density

Earlier in this chapter, we noted that the general term of solid for three-dimensional figures could be misleading, because they could, in many cases, be hollow. But when they are in fact solid, it is possible to talk about the density of the polyhedron. Density is a comparison of an object's mass to its volume. The density of water is 1 gram per cubic centimeter, while the density of honey is 1.33 grams per cubic centimeter, and the density of vegetable oil is a bit less than the density of water, at 0.92 g/cm³. The materials from which you might construct containers also have different densities. The density of aluminum is about 2.7 grams per cubic centimeter, while steel has a density of 8.05 g/cm³ and flexible PVC plastic is about 1.29 g/cm³.

Since $\text{density} = \dfrac{\text{mass}}{\text{volume}}$, if you're asked to find the mass of an object,

$\text{mass} = \text{density} \cdot \text{volume}$.

Remember that mass, the actual amount of matter, is different from weight, although they are often spoken of as though they were the same. Weight depends on the force of gravity acting on the mass. If you travel into space, you may experience weightlessness. If you land on the moon, you will weigh less than you weigh on Earth, but your mass is always the same.

## EXERCISE 14.8

**Mass, weight, volume, density, capacity: all related ideas, similar but not the same. Read carefully, work accurately, and be sure to show units.**

1. A cylinder is 14.5 cm tall and has a diameter of 5.5 cm. If the cylinder is made of aluminum, which has a density of 2.7 grams per cubic centimeter, what is the mass of the cylinder to the nearest gram?

2. If the same cylinder described in question 1 were manufactured from steel, which has a density of 8.05 g/cm³, by how many grams would the weight of the cylinder increase?

3. A can of cola contains 355 mL of cola (or 355 cubic centimeters). That cola has a mass of 394 grams. What is the density of the cola?

4. A can of diet cola also contains 355 mL (or 355 cm³) of soda. The diet cola has a mass

of 355.1 grams. What is the density of the diet cola?

5. What might explain the difference in the densities of these two similar soft drinks in question 3 and 4?

6. A jeweler created a bead that is a sphere with a small cylindrical hole drilled through its center. The diameter of the sphere is 1 cm, and the diameter of the hole is 0.1 cm. The jeweler can create the bead in gold or silver. Gold has a density of 19.32 grams per cubic centimeter, and silver has a density of 10.49 g/cm³. By how many grams will the mass of the gold bead exceed the mass of the silver bead?

# Three-Dimensional Figures

This final review section is focused entirely on three-dimensional figures known as solids. In Chapter 14, you looked at each type separately. These practice questions will give you the opportunity to bring together different shapes and different ideas. Answer all the questions, and try to express your thinking as clearly as you can.

1. Find the number of faces of a polyhedron with 18 edges and 12 vertices.

2. Sketch a possible map for a triangular prism.

3. Find the surface area of hexagonal prism with a perimeter of 12 inches and a height of 3 inches.

4. Find the surface area of a cylinder with a radius of 8 cm and a height of 14 cm.

5. Find the volume of pentagonal prism if the bases are regular pentagons with a perimeter of 40 feet and apothems of 5.5 feet and the height of the prism is 12 feet.

6. Find the volume of a cylinder with a diameter of 28 inches and a height of 30 inches.

7. Find the slant height of a cone with a radius of 36 inches and a height of 77 inches.

8. Find the surface area of a pyramid, to the nearest tenth of a square centimeter, if the base is a regular octagon with a perimeter of 120 cm and an apothem of 18.1 cm and the height of the pyramid is 11 cm.

9. Find the surface area of a cone with a diameter of 18 inches and a height of 12 inches.

10. Find the volume of a square pyramid 4 feet on a side and 6 feet high.

11. Find the volume of a cone with a base circumference of $38\pi$ inches and a height of 21 inches.

12. Find the surface area of a rectangle prism that measures 2 feet by 8 feet by 15 feet.

13. Find the volume of a cone that has a surface area of $216\pi$ square centimeters if the slant height is 6 cm longer than the radius. Leave your answer in terms of $\pi$.

14. Find the height of a triangular pyramid whose base is an equilateral triangle with a perimeter of 12 inches, if the volume of the pyramid is 34 cubic inches.

15. Find the radius of a cylinder with a height of 5 meters and a surface area of $168\pi$ square meters.

16. A square pyramid with a perimeter of 40 cm has a surface area 360 square centimeters. Find its volume.

17. Which has the larger surface area: a cylinder with a diameter of 20 inches and height of 20 inches, or a cone with radius 15 inches and height of 36 inches?

18. The TransAmerica Pyramid, one of the tallest buildings in San Francisco, California, is not a true pyramid because it has "wings" that provide additional space for the narrower upper floors. The underlying pyramid shape has a base area of 65,225 square meters and the building has a height of 260 meters. Find the volume of the building, without its "wings," to the nearest cubic meter.

19. The Coliseum in Rome, Italy, does not have a perfectly circular base, but is actually somewhat elliptical. The area of its base can be calculated nonetheless, using the formula for the area of an ellipse. If the area of the base is $79,515\pi$ square feet and the height is 157 feet, what is the volume of the structure?

20. Iron has a density of approximately 7.9 grams per cubic centimeter. A cylindrical disk with a diameter of 6 cm and a height of 2 cm is cast from iron. The same cylinder, cast in lead, has a mass 192 grams greater. To the nearest tenth, what is the density of lead?

# Posttest

This test is your chance to bring together all the material you have learned through this book. It may ask you to combine ideas in new ways. In the multiple-choice section, choose the best answer for each question. In the free response questions, show all your work, and explain your thinking as clearly as possible.

## PART 1

**Questions 1 to 30 are multiple-choice questions. Choose the best answer for each question.**

1. What transformation would NOT map Triangle I onto Triangle II?

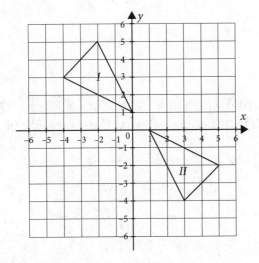

   A) Reflection across the line $y = x$

   B) 180° rotation about the point (½, ½)

   C) Translation 5 units right and 5 units down

   D) Reflection across $y = ½$, followed by reflection across $x = ½$

2. The diagram shows circle _O_ with secants $\overline{PR}$ and $\overline{PV}$ drawn from point _P_ outside the circle. Chord $\overline{RT}$ is also shown. $m\angle QRT = 47°$, $m\,\overset{\frown}{ST} = 74°$, and $m\angle VUT = 103°$. What is the measure of $\angle RPV$?

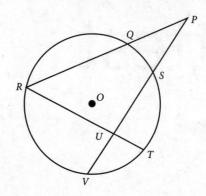

A) 20°

B) 30°

C) 40°

D) 60°

3. In $\triangle ABC$, $\overline{DE}$ is parallel to $\overline{BC}$ and divides $\overline{AB}$ and $\overline{AC}$ in ratio 1:3. $EC = 21$ cm, $DB = 12$ cm, and $BC = 24$ cm. Find the perimeter of $\triangle ADE$ to the nearest centimeter.

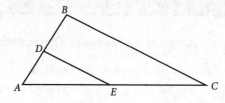

A) 17

B) 19

C) 22

D) 35

4. _PQRT_ is a trapezoid with $\overline{QR} \parallel \overline{PT}$ and $\overline{RS} \parallel \overline{PQ}$. $m\angle Q = 108°$. If $\triangle SRT$ is isosceles with vertex angle $\angle T$, what is the measure of $\angle SRT$?

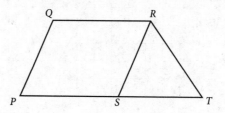

A) 72°

B) 108°

C) 144°

D) 36°

**5.** Which of the following polyhedrons, if sliced parallel to its base, could NOT have a square cross section?

    A) Cube

    B) Rectangular prism

    C) Triangular prism

    D) Square pyramid

**6.** In right triangle $\triangle ABC$, with $\overline{AB} \perp \overline{BC}$, $\sin(\angle C) = \dfrac{8}{17}$. Which of these is also equal to $\dfrac{8}{17}$?

    A) $\sin(\angle A)$

    B) $\sin(\angle B)$

    C) $\cos(\angle A)$

    D) $\cos(\angle C)$

**7.** $\triangle ABC$ is not drawn to scale. Side $\overline{AC}$ is extended through point $C$ to point $D$. $m\angle BCD = 115°$. Which of the following is a true statement?

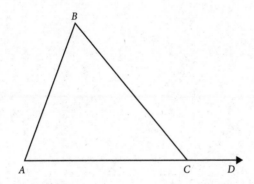

    A) $m\angle ABC = 115°$

    B) $m\angle ACB = m\angle BAC$

    C) $m\angle ABC + m\angle BCA = 115°$

    D) $m\angle ABC + m\angle BAC = 115°$

**8.** In right triangle $\triangle ABC$ with right angle $\angle B$, $\overline{BD}$ is drawn perpendicular to hypotenuse $\overline{AC}$. If $AD = 4$ cm, and $DC = 12$ cm, find the perimeter of $\triangle ABC$ to the nearest tenth of a centimeter.

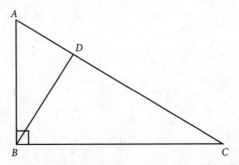

A) 13.9

B) 21.9

C) 37.9

D) 55.4

9. The vertices of $\triangle RST$ have coordinates $R(6,3)$, $S(4,1)$, and $T(0,7)$. What is the perimeter of $\triangle RST$, to the nearest tenth?

A) 10.0

B) 17.2

C) 13.0

D) 14.5

10. If quadrilateral $MNOP$ is a parallelogram, which of these is not sufficient to prove $MNOP$ is a rectangle?

A) $MO = NP$

B) $\overline{MO} \perp \overline{NP}$

C) $\overline{MN} \perp \overline{MP}$

D) $\angle M \cong \angle N$

11. Solve for $x$: $\sin(2x + 7) = \cos(4x - 1)$

A) 4

B) 11.5

C) 14

D) 19

12. In the diagram of circle $O$, $\overline{AB}$ and $\overline{CD}$ intersect at point $E$, $m\angle AEC = 42°$ and $m\overarc{AC} = 51°$. What is $m\overarc{DB}$?

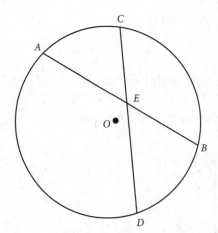

A) 9°

B) 33°

C) 42°

D) 51°

**13.** $\triangle ABC$ is a right triangle with right angle $\angle B$. $\overline{FE} \parallel \overline{AB}$. If $\underline{AD}$, $DB$, $AF$, and $FC$ have the measures shown and $EC = 2\sqrt{3}$, how long is $\overline{DE}$?

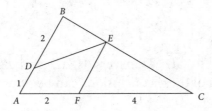

A) 3

B) $\sqrt{3}$

C) 7

D) $\sqrt{7}$

**14.** The Point Arena Light in northern California is a 115-foot concrete tower, built on a rocky point 40 feet above sea level, placing the light 155 feet above sea level. The light can be seen more than 25 miles out to sea. If a ship is 1 mile out to sea, what is the measurement of the angle of elevation to the top of the lighthouse, to the nearest degree? (1 mile = 5,280 feet)

A) 1°

B) 2°

C) 42°

D) 89

**15.** The coordinates of the endpoints of $\overline{XY}$ are $X(1, 9)$ and $Y(4, -1)$. Find the coordinates of point $P$ that divides $\overline{XY}$ into two segments whose lengths are in ratio $XP{:}PY = 2{:}3$.

A) (1, 5)

B) (2.2, 7.4)

C) (2.2, 5)

D) (2.2, 0.6)

**16.** The image of $\triangle RST$ under a reflection is $\triangle R'S'T'$. The pre-image and image are shown. Which of the following equations represents the line of reflection?

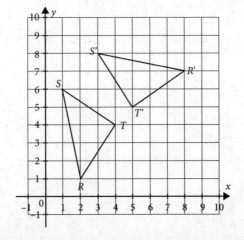

A) $y = x$

B) $y = -x$

C) $y = 9 - x$

D) $y = x - 9$

17. Line $\overleftrightarrow{AB}$ is parallel to $\overleftrightarrow{CD}$. Transversals $\overleftrightarrow{AD}$ and $\overleftrightarrow{BC}$ intersect at point $E$ as shown. Which of these conclusions can be drawn?

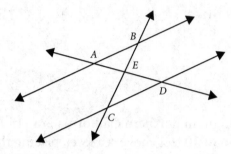

A) $\triangle ABE \cong \triangle CED$

B) $\triangle ABE \cong \triangle DEC$

C) $\triangle ABE \sim \triangle CED$

D) $\triangle ABE \sim \triangle DCE$

Use the diagram below for questions 18 and 19.

18. In $\triangle RST$, $RS = 21$ cm, $RT = 16$ cm, and m$\angle R = 80°$. What is the area of $\triangle RST$ to the nearest tenth of a square centimeter?

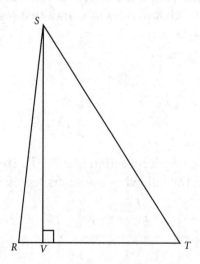

A) 165.4

B) 168.0

C) 29.2

D) 952.8

19. Using the figure in question 18, find the height *SV* of △*RST* to the nearest tenth of a centimeter.

    A) 10.3

    B) 12.0

    C) 18.3

    D) 20.7

20. A regular octagon is rotated about its center. Which of these rotations will NOT map the octagon onto itself?

    A) 45° clockwise

    B) 60° counterclockwise

    C) 90° clockwise

    D) 180° counterclockwise

21. A right triangle with legs of 18 cm and 24 cm is rotated about a line to create a cone with a volume of 2,592π cubic centimeters. About which of these lines is the triangle rotated to produce a cone with that volume?

    A) The 18 cm leg

    B) The 24 cm leg

    C) The hypotenuse

    D) The altitude to the hypotenuse

22. Line segment *l* is shown in the diagram. Which of these points does not lie on the perpendicular bisector of line segment *l*?

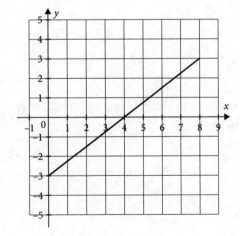

    A) (4, 0)

    B) (1, 4)

    C) (7, −4)

    D) (3, 1)

| State | Population Density (people per square mile) | Land Area (square miles) |
|---|---|---|
| New Hampshire | 148 | 8,953 |
| Indiana | 184 | 35,826 |
| Louisiana | 108 | 43,204 |
| Georgia | 177 | 57,513 |

**23.** The table above shows four states with similar population density, in people per square mile, and their land area in square miles. Rank the states according to population, from smallest to largest.

A) Indiana, Georgia, New Hampshire, Louisiana

B) Louisiana, New Hampshire, Georgia, Indiana

C) Georgia, Indiana, Louisiana, New Hampshire

D) New Hampshire, Louisiana, Indiana, Georgia

**24.** Two triangles are shown below. Markings on the diagram show that $AB \cong XZ$ and $\angle C \cong \angle Y$. Which of the following statements must be shown to be true in order to prove $\triangle ABC \cong \triangle XZY$?

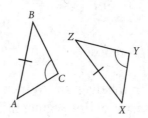

A) $\angle A \cong \angle X$

B) $\overline{AC} \cong \overline{XY}$

C) $\overline{BC} \cong \overline{ZY}$

D) $\angle A \cong \angle Z$

**25.** Rhombus $PQRS$ has diagonals $\overline{PR}$ and $\overline{QS}$ that intersect at point $T$. $PR = 24$ cm and $QS = 9$ cm. Find the perimeter of $PQRS$ to the nearest tenth of a centimeter.

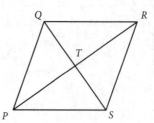

A) 12.8

B) 31.2

C) 51.2

D) 60.0

26. Which of these is the equation of a circle of radius 7 centered at the point $(3, -4)$?

   A) $(x + 3)^2 + (y - 4)^2 = 49$

   B) $(x - 3)^2 + (y + 4)^2 = 7$

   C) $(x + 3)^2 + (y - 4)^2 = 7$

   D) $(x - 3)^2 + (y + 4)^2 = 49$

27. In $\triangle ABC$, $m\angle A = 45°$ and $m\angle B = 110°$. In $\triangle RST$, $m\angle R = 45°$ and $m\angle S = 25°$. Which statement about these two triangles is correct?

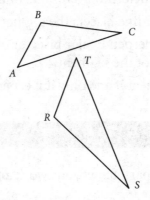

   A) $\triangle ABC \sim \triangle RST$

   B) $\triangle ABC \sim \triangle RTS$

   C) $\triangle ABC \sim \triangle TRS$

   D) $\triangle ABC$ and $\triangle RST$ have no similarity relationship.

28. Circle $O$ has a diameter of 30 cm. $\angle AOB$ is a central angle, and the sector defined by $\angle AOB$ has an area of $20\pi$. What is the length of arc $\overset{\frown}{AB}$ to the nearest tenth of a centimeter?

   A) 2.1

   B) 8.4

   C) 32.0

   D) 62.8

Use the figure below for questions 29 and 30.

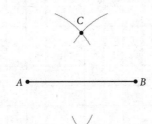

29. The diagram above shows part of the work of constructing the perpendicular bisector of segment $AB$. What is the next necessary step in the construction?

A) Draw $\overline{AC}$

B) Draw $\overline{CD}$

C) Draw an arc centered at $C$, passing through $D$

D) Bisect $\overline{CD}$

30. Which of these statements provides the rationale for the construction of the perpendicular bisector of segment $AB$?

A) Two distinct points determine a unique line.

B) Every perpendicular line bisects the segment to which it is drawn.

C) Every point on the perpendicular bisector of a segment is equidistant from the endpoints of the segment.

D) A line bisects a segment if and only if it is perpendicular to the segment.

## PART 2

Answer all questions in this part. Show your work and explain your reasoning.

31. The points $A(1,1)$, $B(3,2)$, $C(3,5)$, and $D(1, 4)$ are the vertices of a quadrilateral. The quadrilateral is transformed by reflection across the $x$-axis, followed by a dilation by a factor of 2 centered at the origin. Draw the original quadrilateral and its image under these transformations.

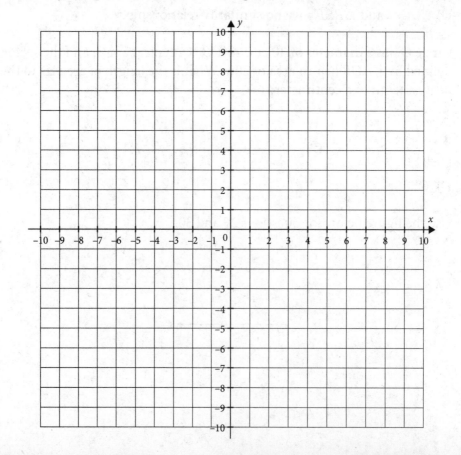

**32.** Use compass and straightedge to construct a regular hexagon inscribed in the circle below.

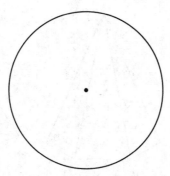

**33.** *ABCD* is an isosceles trapezoid and *F* is the midpoint of $\overline{BC}$. $\angle EAD \cong$ $\angle EDA$. Prove that $\triangle AFE \cong \triangle DFE$.

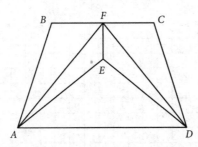

**34.** Parallelogram *ABCD* is drawn in the coordinate plane with the vertices shown. Use the coordinates given to prove or disprove the claim: The line segment connecting the midpoints of opposite sides of parallelogram *ABCD* bisects both diagonals.

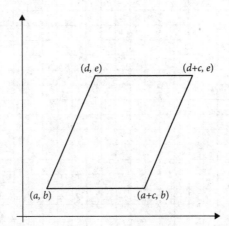

**35.** The diagram shows a cone and two pyramids, one with a regular hexagon as its base, and one with a square base. All three figures have the same height and the same radius. (The radius of a pyramid is the distance from the center of the regular polygon to a vertex.) Which of these figures has the greatest volume? Explain your reasoning.

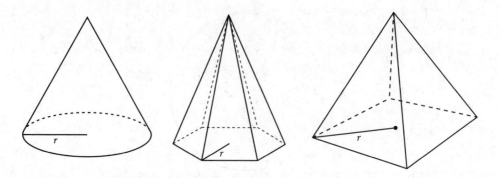

**36.** Find the center and radius of the circle whose equation is $x^2 + 12x + 36 + y^2 - 6y + 9 = 81$. Sketch its graph.

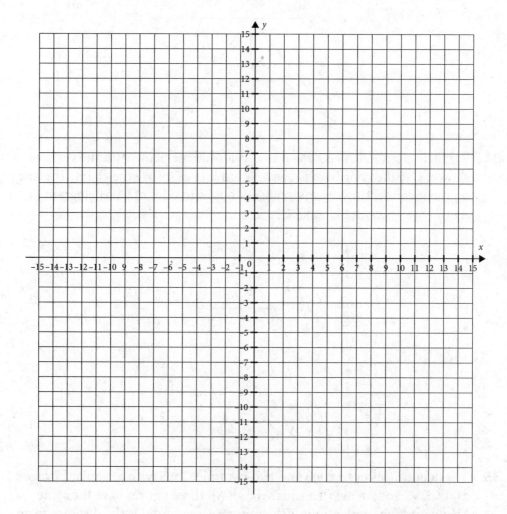

**37.** The Johnson family's home is built on a trapezoidal lot as shown in the diagram. The house is a rectangle, and a trapezoidal deck was added to one side of the house. A paved driveway in the shape of a parallelogram leads from the edge of the property to a small rectangular garage.

There is a square garden shed in one corner of the property. The Johnsons want to add a pool to their property. They would like the pool to be as large as possible, but it cannot come in contact with any existing structure or the edge of the property. Give the Johnsons your best recommendation for the shape, size, and location of the pool. Use the coordinate system to locate and measure. Explain the reasoning behind your recommendation.

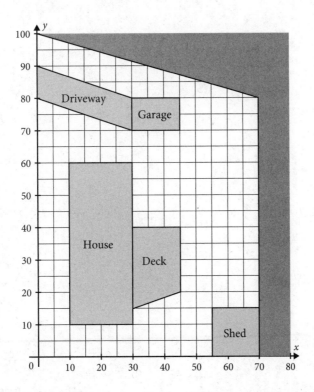

**38.** The Pentagon, the building that houses the headquarters of the U.S. Department of Defense, is a regular pentagon 281 meters on a side. The central courtyard, also a regular pentagon, has an area of 20,000 square meters. What is the area of the building itself to the nearest square meter? How long is the side of the courtyard in meters?

# Glossary

## Symbols

≅    "Is congruent to"; denotes angles have equal measure, segments have equal length, or triangles (or other polygons) have congruent corresponding angles and congruent corresponding sides.

$\pi$    Pi, a number defined as the ratio of the circumference of a circle to its diameter. Approximately 3.14159….

~    "Is similar to" (In logic, the negation. See ~$p$).

~$p$    The negation or opposite of statement $p$. "not $p$."

=    "Is equal to"; used when talking about numbers. If two angles are congruent ($\angle A \cong \angle B$), the measures of those angles are equal (m$\angle A$ = m$\angle B$).

## Vocabulary

**AA**    Angle-angle; denotes the most common way of proving triangles similar: two angles of one triangle are congruent to the two corresponding angles of the other.

**AAS**    Angle-angle-side.

**Absolute value**    The value of a number without regard to its sign.

**Acute triangle**    A triangle containing three angles all less than 90°.

**Adjacent angles**    Two angles which share a vertex and share one side and do not overlap.

**Adjacent side**    In a right triangle, the leg that meets the hypotenuse to form an acute angle is the angle's adjacent side.

**Alternate exterior angles**    When two lines are cut by a transversal, an angle on one side of the transversal at the intersection with one line and an angle on the other side of the transversal at the intersection with the other line are alternate exterior angles if neither angle is in the space between the lines.

**Alternate interior angles**   When two lines are cut by a transversal, an angle on one side of the transversal at the intersection with one line and an angle on the other side of the transversal at the intersection with the other line are alternate interior angles if both angles are located in the space between the lines.

**Altitude**   A line segment from a vertex of a triangle perpendicular to the opposite side. The length of that segment, sometimes called the *height*. In a parallelogram, a segment parallel to both of two opposite sides.

**Ambiguous case**   A situation that arises when using the Law of Sines to find a missing angle in which two measurements, supplements of one another, are possible for the angle.

**Angle**   The figure formed by two rays with the same endpoint. The common endpoint is the *vertex* of the angle, and the two rays are the *sides* of the angle.

**Angle bisector**   A ray, line, or segment that passes through the vertex of an angle and divides it into two angles of equal measure.

**Angle Bisector theorem**   A segment that bisects one angle of a triangle divides the opposite side into two segments proportional to their adjacent sides.

**Angle formed by tangent and chord**   An angle whose vertex is on the circle at the point of tangency and whose sides are the tangent line and a chord drawn to the point of tangency.

**Angle formed by tangent and radius**   An angle with its vertex at the point of tangency whose sides are the tangent and a radius drawn to the point of tangency.

**Angle formed by tangent and secant**   An angle whose vertex is a point outside the circle, formed when a tangent and a secant are drawn to the same circle from that external point.

**Angle formed by two secants**   An angle whose vertex is a point outside the circle, formed when two secants are drawn to the same circle from that external point.

**Angle formed by two tangents**   An angle whose vertex is a point outside the circle, formed when two tangents drawn to the same circle from that external point.

**Angle of depression**   The angle formed by a horizontal line and a line of sight to some point or object below.

**Angle of elevation**   The angle formed by a horizontal line and a line of sight to some point or object above.

**Apothem**   A line segment from the center of a regular polygon perpendicular to one edge, or the length of such a segment.

**Arc**   A portion of a circle between two points on the circle.

**Arc length**   A fraction of the circumference of the circle between two points.

**Arc measure**   The number of degrees covered by an arc; the amount of rotation between the radii of the central angle that intercepts the arc.

**Area**   The measurement of a section of the plane enclosed within a polygon or other figure.

**Area of sector**   The portion of the area of a circle enclosed by two radii and their intercepted arc.

**Argument**   A set of statements and reasons meant to prove the truth of a statement.

**ASA**   Angle-side-angle; a shortcut for showing triangles congruent by proving two angles and the included side of one triangle congruent to two angles and the included side of another.

**Base**   A side of a polygon to which an altitude is drawn; the noncongruent side of an isosceles triangle; in a prism or cone, one of two parallel faces.

**Base Angle Converse theorem**   A theorem that says that if two angles of a triangle are congruent, the sides opposite those angles are congruent.

**Base Angle theorem**   A theorem that says that if two sides of a triangle are congruent, the angles opposite those sides are congruent.

**Base angles**   In an isosceles triangle, the angles at each end of the noncongruent side. In a trapezoid, the angles at each end of either of the parallel sides.

**Biconditional**   A statement that is the conjunction of a conditional statement and its converse: $(p{\rightarrow}q){\wedge}(q{\rightarrow}p)$, also written as $p{\leftrightarrow}q$ and read as "$p$ if and only if $q$."

**Bisect**   To cut into two pieces of equal size.

**Bisector**   A line, ray, or segment that intersects a line segment at its midpoint or divides an angle into two angles of equal measure.

**Cartesian plane**   A plane (or flat surface) on which a rectangular coordinate system has been established to allow every point in the place to be located by an ordered pair of coordinates $(x, y)$ with respect to a point called the *origin*. The first coordinate describes horizontal movement, and the second describes vertical movement.

**Center (of circle)**   A fixed point around which the points of a circle are arranged, all at the same distance from the center point.

**Center (of dilation)**   A fixed point from which rays project through the vertices of a pre-image and define the vertices of the image.

**Central angle (circle)**   An angle formed by two radii whose vertex is the center of the circle.

**Central angle (polygon)**   An angle with its vertex at the center of a regular polygon, formed by radii from the center to the vertices at each end of a side.

**Centroid**   The point at which the bisectors of the angles of a triangle intersect; a point of concurrence.

**Chord**   A line segment with endpoints on a circle.

**Circle**   Where a set of all points in a plane are at a fixed distance, called the *radius*, from a given point called the *center*.

**Circumcenter**   The point at which the perpendicular bisectors of the sides of a triangle intersect; a point of concurrence.

**Circumference**   The distance around a circle.

**Circumscribed polygon**   A polygon whose sides are all tangent to a circle.

**Coefficient**   One of the factors being multiplied in a term, usually used to denote a numerical or constant factor that is multiplied by a variable.

**Common external tangent**   A line that touches two circles but does not cross the line segment that connects the centers.

**Common internal tangent**   A line that touches two circles at an angle that allows it to cross the line segment that connects the centers of the circles.

**Compass**   A tool for scribing circles and arcs, or measuring and duplicating distances.

**Complementary**   Two angles are complementary if their measures add to 90°. Each is the complement of the other.

**Complementary identities**   Statements relating trig functions of an angle and its complement; for example, if $\angle A$ and $\angle B$ are complementary, $\sin(\angle A) = \cos(\angle B)$.

**Composition**   The creation of a new function or operation by having two or more functions or operations performed in succession.

**Composition transformations**   Applying one transformation, and then applying another transformation to the result.

**Compound statement**   A statement created from two (or more) simple statements combined with and, or, or if-then.

**Concave**   Not convex.

**Concentric circles**   Circles of different radii which have the same center.

**Conclusion**   In a conditional statement, the portion that follows "then." The idea that follows from the hypothesis.

**Concurrent**   Three or more lines intersecting at a single point are concurrent.

**Conditional**   An if-then statement; the statement that a certain hypothesis implies a certain conclusion.

**Cone**   A three-dimensional figure with a circle for its base and a lateral surface that tapers to a point.

**Congruence**   A correspondence between two polygons in which all corresponding angles have equal measures and all corresponding sides have equal lengths.

**Congruent**   For angles or segments, having the same measurement. For polygons, having a congruence relationship. (See also *Congruence*.)

**Conic section**   A figure that results from slicing a cone: circle, ellipse, hyperbola, parabola.

**Conjunction**   A compound statement in which two simple statements are connected by the word "and."

**Consecutive angles of parallelogram**   The angles at each end of any side of a parallelogram.

**Consecutive interior angles**   When two lines are cut by a transversal, two angles on the same side of the transversal, one at each intersection of the transversal and a line, and both falling between the lines. Also called *same-side interior angles*.

**Construct**   Create a drawing using only compass and straightedge.

**Contrapositive**   A statement logically equivalent to a conditional $p{\rightarrow}q$. The contrapositive has the form ${\sim}q{\rightarrow}{\sim}p$.

**Converse**   A statement related to the conditional $p{\rightarrow}q$ in which the hypothesis and conclusion are reversed: $q{\rightarrow}p$.

**Convex**   A polygon is convex if all its diagonals fall within the polygon.

**Convex polyhedron**   A polyhedron in which any line connecting two points on the surface of the polyhedron is inside the polyhedron.

**Coordinate plane**   A plane or surface on which a system has been established to allow points to be located by an ordered pair of numbers. The most commonly used system of rectangular coordinates use a pair of number lines, called *axes*, one horizontal and one vertical, that intersect at their zeros. A point is located by a pair of numbers that direct the horizontal and vertical movement from the zero, or origin, to the desired point.

**Coordinate proof**   An effort to demonstrate the truth of a geometric statement by creating a diagram on a coordinate plane that represents the hypothesis in general terms rather than specific coordinates, and then using coordinate formulas—like the distance formula, midpoint formula, and slope formula—to show that the conclusion will be true.

**Coplanar**   On the same plane.

**Corresponding angles**   When two lines are cut by a transversal, each intersection has a cluster of four angles. The angles in one cluster correspond to the angles in the other cluster according to their position in the cluster: upper left to upper left, upper right to upper right, and so on.

**Corresponding angles**   In congruent polygons or similar polygons, the angles that align and are congruent are corresponding angles.

**cos$^{-1}$( )**   "The angle whose cosine is . . . "; the inverse of the cosine.

**Cosecant**   In a right triangle, the ratio of the length of the hypotenuse to the length of the opposite side; the reciprocal of the sine.

**Cosine**   In a right triangle, the ratio of the side adjacent to an acute angle to the hypotenuse.

**Cotangent**   In a right triangle, the ratio of the side adjacent to an acute angle to the side opposite that angle; the reciprocal of the tangent.

**CPCTC**   "Corresponding parts of congruent triangles are congruent."

**Cross-multiply**   To multiply the means of a proportion and set the product equal to the product of the extremes. To convert $\frac{a}{b} = \frac{c}{d}$ to $ad = bc$.

**Cross section**   The two-dimensional figure that results when a three-dimensional object is cut through by a plane.

**Cube**   A regular hexahedron; a rectangular prism in which all six faces are squares.

**Cylinder**   A three-dimensional figure with two circular bases and a lateral surface that unrolls to a parallelogram.

**Decagon**   A 10-sided polygon.

**Degree**   A unit for measuring rotation. A full rotation is divided into 360 degrees. Angles and arcs can be measured by the number of degrees of rotation they capture.

**Density**   A comparison of the mass of an object to its volume.

**Detachment**   A logical structure known to be a valid argument. $[(p{\rightarrow}q){\wedge}p]{\rightarrow}q$.

**Diagonal of a polygon**   A line segment connecting two vertices of the polygon that are not endpoints of the same side.

**Diameter**   A line segment with endpoints on the circle that passes through the center; the longest chord of a circle.

**Dilation**   A transformation which enlarges or reduces a pre-image.

**Directrix**   A line that, along with the focus point, determines the shape of a parabola.

**Disjunction**   A compound statement in which two simple statements are joined by the word "or."

**Distance between two points**   The shortest path between two points. On a number line, the absolute value of the difference of the coordinates. On the coordinate place, the length of the oblique line connecting the points.

**Distance formula**  A variant of the Pythagorean theorem using coordinates of two points in the plane, resulting in the distance between the two points:

$$d = \sqrt{(x_2 - x_1)^2 + (y_2 - y_1)^2} \; .$$

**Dodecagon**  A 12-sided polygon.

**Edge**  The line segment at which two faces of a polyhedron intersect.

**Ellipse**  An oval shape created when a cone is sliced at an angle.

**Endpoints**  Two points in the plane that define a line segment or an arc.

**Enlargement**  A dilation that results in an image that is larger than the pre-image.

**Equation of a line**  A mathematical statement that relates the $x$- and $y$-coordinates of the points on a line. An equation of the form $y = mx + b$ or $y - y_1 = m(x - x_1)$ or $ax + by = c$.

**Equiangular**  Having all interior angles of equal measure.

**Equilateral**  Having all sides of equal length.

**Equilateral triangle**  A triangle with three sides of equal length.

**Equivalent**  In logic, two statements are equivalent if they always have the same truth value.

**Exclusive or**  A variant of a disjunction which is true only when exactly one of the simple statements is true. One or the other but not both.

**Exterior angle**  An angle formed by extending one side of a polygon through one vertex that is its endpoint.

**Exterior Angle Inequality theorem**  The measure of an exterior angle of a triangle is greater than the measure of either of the two nonadjacent angles.

**Exterior Angle theorem**  The measure of an exterior angle of a triangle is equal to the sum of the measures of the two nonadjacent angles.

**External segment of tangent or secant**  The portion of a tangent or secant between the point from which it is drawn and the point at which it first intersects the circle.

**Externally tangent circles**  Two circles which touch at a single point but have no interior points in common.

**Extremes**  In a proportion, the first and last quantities are called the extremes.

$$\text{Extreme: mean} = \text{mean: extreme} \quad \text{or} \quad \frac{\text{extreme}}{\text{mean}} = \frac{\text{mean}}{\text{extreme}}.$$

**Face**  A portion of a plane, bounded by a polygon, which combines with others to form a polyhedron.

**Focus (focii, focal points)**  In a conic section, a point or pair of points whose location and relationship to one another affect the shape of the graph.

**Function**  A rule defined by an equation, graph, or list of points that carries inputs to outputs, with each input have only one output.

**General form**  A form of the linear equation, with the variable terms on one side equal to a constant, and all coefficients integers.

**Geometric mean**  A number which occupies both mean positions in a proportion.

**Heptagon**  A seven-sided polygon.

**Hexagon**  A six-sided polygon.

**Hinge Converse theorem**  If two sides of one triangle are congruent to the corresponding sides of another triangle but the included angles are not congruent, the triangle with the shorter third side has the smaller included angle.

**Hinge theorem**    If two sides of one triangle are congruent to the corresponding sides of another triangle but the included angles are not congruent, the triangle with the smaller included angle has the shorter third side.

**Horizontal line**    A line with a slope of zero; a line that never rises.

**Hyperbola**    A conic section created by slicing vertically through both pieces of a double cone; a graph that resembles a set of wings curving away from one another.

**Hypotenuse**    The longest side of a right triangle, which is opposite the right angle.

**Hypothesis**    The "if" portion of a conditional statement; the idea that leads to the conclusion.

**Identity**    In algebra and trigonometry, an equation that is true for any value of the variable.

**Identity property**    In logic, the postulate that says any quantity is equal to itself or any polygon is congruent to itself.

**Iff**    "If and only if"; a convenient way to read or write a biconditional $p \leftrightarrow q$; "$p$ iff q" is equivalent to "if $p$ then $q$ and if $q$ then $p$."

**If-then**    The structure of the conditional statement $p \rightarrow q$ or if $p$ then $q$.

**Image**    The result of a transformation. A transformation carries a pre-image to an image.

**Incenter**    The point at which the three angle bisectors in a triangle intersect.

**Inscribed angle**    An angle whose radius is a point on a circle and whose sides are chords.

**Inscribed polygon**    A polygon all of whose vertices are points on the circle.

**Integer**    A positive or negative whole number

**Interior angle of a polygon**    An angle formed when two sides of a polygon meet at a vertex.

**Internally tangent circles**    Two circles which touch at a single point with one circle inside the other.

**Inverse**    The opposite or undoing; the function or operation that sends the output of a function or operation back to the input from which it came.

**Isosceles triangle**    A triangle with two sides of equal length.

**Isosceles trapezoid**    A quadrilateral with one pair of parallel sides in which the nonparallel sides are the same length.

**Lateral area**    The area of the polygons that surround the base(s) of a prism or pyramid (but not including the bases themselves). The curved surface of a cylinder or cone.

**Law of Cosines**    A rule that relates the lengths of sides of any triangle and the cosine of one angle of that triangle: $c^2 = a^2 + b^2 - 2ab\cos(\angle C)$.

**Law of Sines**    A rule that states that the lengths of the sides of a triangle are proportional to the sines of the opposite angles: $\dfrac{a}{\sin(\angle A)} = \dfrac{b}{\sin(\angle B)} = \dfrac{c}{\sin(\angle C)}$.

**Leg of a right triangle**    One of the two sides that form the right angle.

**Leg of an isosceles triangle** One of the two congruent sides.

**Line** An undefined term, described as a one-dimensional object that has length but no width or height, made up of infinitely many points side by side, without bends or curves.

**Line segment** A part of a line that includes two points on the line, called *endpoints*, and all the points on the line between them.

**Linear pair** A pair of adjacent angles whose exterior sides form a straight angle (a line).

**Major arc** An arc larger than a semicircle, measuring more than 180°.

**Major axis** The longer of the two axes of an ellipse.

**Mass** The amount of matter in an object.

**Means** Two of the four quantities in a simple proportion, specifically, the two middle quantities. Extreme: mean = mean:extreme.

**Means-Extremes property** In any proportion, the product of the means equals the product of the extremes.

**Median** A line segment that connects a vertex of a triangle to the midpoint of the opposite side.

**Midpoint** A point on a line segment that divides the segment into two segments of equal length; a point equidistant from the endpoints of a segment.

**Midsegment** A line segment that connects the midpoints of two sides of a triangle.

**Midsegment theorem** The midsegment of a triangle is parallel to the third side and half as long.

**Minor arc** An arc of less than 180°.

**Minor axis** The shorter of the two axes of an ellipse.

**Negation** A statement that is the opposite or contradiction of another.

**Negative reciprocal** Two numbers are negative reciprocals if their product is –1. The negative reciprocal of $\frac{a}{b}$ is $-\frac{b}{a}$.

**Net** A drawing in the plane that decomposes a three-dimensional object into a figure that could be folded into the original object.

**Nonagon** A polygon with nine sides.

**Non-convex** A polygon that contains one or more interior angles that contain more than 180°, causing the polygon to have diagonals that fall outside the polygon.

**Number line** A line to which a measuring system has been attached to show the relative positions of the real numbers.

**Oblique prism** A prism in which the parallel bases are not directly above one another, causing the prism to slant.

**Obtuse triangle** A triangle containing one angle greater than 90°.

**Octagon** A polygon with eight sides.

**Opposite rays** Two rays with the same endpoint, but moving in opposite directions, so that the result looks like a line.

**Opposite side** In a right triangle, the leg across from one of the acute angles; the side that is not a side of the angle.

**Orientation** The angle of an object with respect to compass directions or the coordinate axes. A line, for example, may be angled 45° clockwise from the $y$-axis. If that line is reflected, its orientation will change to 45° counterclockwise.

**Origin**   The point at which the $x$-axis and $y$-axis cross, the zero of both axes.

**Orthocenter**   The point at which the three altitudes of a triangle intersect.

$p \land q$   "$p$ and $q$"; a conjunction or "and" statement.

$p \lor q$   "$p$ or $q$"; a disjunction or "or" statement.

$p \rightarrow q$   "If $p$, then $q$"; a conditional statement.

**Parabola**   A cup-shaped curve that is the graph of a quadratic equation; the set of all points equidistant from a focus point and a line called the *directrix*; the cross section created when a cone is sliced from lateral side to base.

**Parallel lines**   Two lines on the same plane that never intersect.

**Parallelogram**   A quadrilateral in which both pairs of opposite sides are parallel.

**Parameter**   A characteristic of a system or operation; a symbol that acts as a placeholder in the general form and is replaced by a fixed value in a specific example; for example, $m$ and $b$ in the slope-intercept form of a line, $y = mx + b$, take values for a specific line, like $y = 3x + 1$, but then remain fixed, unlike the variables $x$ and $y$.

**Pentagon**   A polygon with five sides.

**Perfect square**   A number which is the result of multiplying an integer by itself; also used in algebra to denote a trinomial that is the result of multiplying a binomial times itself.

**Perimeter**   The distance around the edges of a polygon; the sum of the lengths of the sides.

**Perpendicular lines**   Two lines that intersect at right angles.

**Plane**   An undefined term, used to denote a flat surface which has infinite length and infinite width but no height.

**Plane geometry**   That portion of the study of geometry that focuses on two-dimensional figures.

**Point**   An undefined term, often represented by a dot, but which theoretically has no length, no width, and no height; a point has location but takes up no space.

**Point of tangency**   The single point on a circle at which a tangent line intersects the circle.

**Point-slope form**   An arrangement of the general equation of a line that allows the specific equation to be written by specifying the slope, $m$, and the coordinates of one point $(x_1, y_1)$ on the line; $y - y_1 = m(x - x_1)$.

**Polygon**   A plane figure whose sides are line segments that intersect only at their endpoints.

**Polyhedron**   A three-dimensional figure formed by polygons that meet at their edges.

**Postulate**   A statement that is accepted by agreement, without proof; a fundamental idea that experience strongly suggests is true.

**Pre-image**   A point, line, segment, or other figure that is to be subjected to a transformation, the result of which is called the *image*.

**Prism**   A polyhedron with bases that are two congruent polygons parallel to one another, and a lateral surface made up of parallelograms connecting the bases.

**Proof**   An explanation that carefully justifies a claim, giving evidence to support the statement and explanations of why those facts should be accepted as true.

**Proportion**   Two equal ratios; a statement that communicates a relationship amongst the sides of similar polygons.

**Pyramid**   A polyhedron with one polygonal base, surrounded by triangles that meet at a point.

**Pythagorean Converse theorem**   If the sum of the squares of the two shorter sides of a triangle is less than the square of the longest side, the triangle is an obtuse triangle; if that sum is more than the square of the longest side, the triangle is an acute triangle.

**Pythagorean identities**   A group of statements about relationships among trigonometric ratios that are derived from the Pythagorean theorem; $\sin^2(\theta) + \cos^2(\theta) = 1$ and variants.

**Pythagorean theorem**   The square of the hypotenuse of a right triangle is equal to the sum of the squares of the other two sides; if the legs of a right triangle measure $a$ units and $b$ units and the hypotenuse measures $c$ units, then $a^2 + b^2 = c^2$.

**Pythagorean triple**   A set of three integers which fit the Pythagorean theorem relationship.

**$q \rightarrow p$**   "If $q$, then $p$"; the converse of the conditional statement $p \rightarrow q$.

**Quadrant**   A portion of the coordinate plane; one of the four sections created by the intersecting axes. The quadrants are numbered I, II, III, IV, beginning at the upper right and moving counterclockwise.

**Quadrilateral**   A polygon with four sides.

**Radian**   A measurement equal to the size of a central angle that intercepts an arc whose length is equal to the length of the radius.

**Radius of a circle**   A fixed distance from a given point called the *center* that defines the size of a circle; a line segment from the center of a circle to a point on the circle.

**Radius of a polygon**   In a regular polygon, a line segment that connects the center to a vertex, or the length of such a segment,

**Ratio**   A comparison of two numbers by division.

**Ray**   A part of a line that includes one point on the line, called the *endpoint*, and all the points on the line to one side of the endpoint.

**Reciprocal**   For any nonzero number, the result of 1 divided by the number, so that the number and its reciprocal multiply to 1.

**Reciprocal identities**   In trigonometry, a group of statements that connect two trigonometric ratios by stating that for any angle, one is the reciprocal to the other; for example, $\csc(\theta) = \dfrac{1}{\sin(\theta)}$.

**Rectangle**   A parallelogram with a right angle.

**Reduction**   A type of dilation that results in an image smaller than the pre-image.

**Reflection**   A transformation in which every point is mapped to a point on the opposite side of a reflecting line, so that the reflecting line is the perpendicular bisector of a line segment connecting the point and its image.

**Reflexive**    A property which says that any number is equal to itself or that any polygon is congruent to itself.

**Regular**    A polygon is regular if it is equilateral and equiangular.

**Rhombus**    A parallelogram with four sides of equal length.

**Right angle**    An angle that measures 90°.

**Right prism**    A prism in which the parallel bases are directly above one another, so that the lateral faces make right angles with the bases.

**Right triangle**    A triangle containing one right angle.

**Rigid transformation**    A transformation that preserves length and angle measure, resulting in an image congruent to the pre-image.

**Rotation**    A rigid transformation in which each point of the pre-image moves along one of several concentric circles for the same number of degrees; two reflections over intersecting lines.

**Same-side interior angles**    If two lines are cut by a transversal, two angles between the two lines, on the same side of the transversal, one with vertex on the first line and one with vertex on the second line.

**SAS**    Side-angle-side; a method for proving two triangles congruent by showing that two sides and the angle included between them in one triangle are congruent to the corresponding parts of the other triangle.

**Scale factor**    The ratio of corresponding sides of two similar triangles; the size of the enlargement or reduction in a dilation.

**Scalene**    A triangle in which all three sides have different lengths.

**Secant of a circle**    A line that intersects a circle in two distinct points; a line that contains a chord.

**Secant of a triangle**    In right triangle trigonometry, the ratio of the length of the hypotenuse to the length of the adjacent side; the reciprocal of the cosine.

**Semicircle**    Half a circle; an arc of 180°.

**Sides of an angle**    Two rays with a common endpoint.

**Side-Splitter theorem**    A line parallel to one side of a triangle divides the other sides proportionally.

**Similar**    Having the same shape but different sizes; having corresponding angles congruent and corresponding sides in proportion.

**$\sin^{-1}()$**    "The angle whose sine is"; the inverse of the sine function.

**Sine**    A trigonometric relationship originally defined in a right triangle as the ratio of the length of the side opposite an acute angle to the length of the hypotenuse.

**Skew lines**    Lines which never intersect but do not lie in the same plane.

**Slant height**    In a pyramid, the altitude of the triangle that forms a lateral face; in a cone, the distance from the point at the top of the cone to the edge of the circular base along the curved lateral surface.

**Slope of a line**    The ratio of the vertical change, or rise, to the horizontal change, or run, that describes the angle of rise or fall of a line.

**Slope-intercept form**    An arrangement of the equation of a line that makes the slope, $m$, and the $y$-intercept, $b$, obvious; $y = mx + b$.

**Solve a triangle**    To find the lengths of any sides not yet known, and the measures of any unknown angles, using methods of trigonometry and geometry.

**Special right triangle**    A triangle from either of two classes of right triangles: the 45°-45°-90° triangle and the 30°-60°-90 triangle. The sides of the 45°-45°-90° triangle are $x$, $x$, and $x\sqrt{2}$. The sides of the 30°-60°-90 triangle are $\frac{1}{2}h$, $\frac{1}{2}h\sqrt{3}$, and $h$.

**Sphere**    A three-dimensional round figure in which every point on the surface is the same distance from the center.

**Square**    A parallelogram with four congruent sides and four right angles; a parallelogram that is both a rhombus and a rectangle.

**SSS**    Side-side-side; a method of proving that two triangles are congruent by showing that all three sides of one triangle are congruent to the corresponding sides of the other.

**Standard form**    The commonly used or characteristic arrangement of the equation of a particular type of equation.

**Statement**    A sentence that can be labeled true or false.

**Straightedge**    A tool to aid in drawing lines, but lacking the measuring scale common on rulers.

**Substitution property**    If $x = y$, then $x$ can be substituted in for $y$ in any equation, and $y$ can be substituted for $x$ in any equation.

**Supplementary**    Two angles whose measures total 180° are supplementary.

**Surface area**    In a three-dimensional figure, the total of the areas of the polygons and/or curved surfaces that form the figure.

**Syllogism**    A pattern of reasoning that links several conditional statements, to conclude that the hypothesis of the first implies the conclusion of the last.

**Symmetry**    A regularity in the shape of an object so that the reflection of the shape across a line within the shape, or the rotation of the shape about a point within the shape, carries it onto itself.

**tan$^{-1}$()**    "The angle whose tangent is"; the inverse of the tangent function.

**Tangent to a circle**    A line that touches the circle at only one point.

**Tangent of a triangle**    The ratio of the side opposite one of the acute angles of a right triangle to the side adjacent to that angle.

**Tautology**    A statement that is always true, without regard to the truth values of the simple statements that combine to form it.

**Tetrahedron**    A polyhedron with four faces.

**Theorem**    A statement whose truth is proven.

**Third Angle theorem**    If two angles of one triangle are congruent to two angles of another triangle, the third angles are congruent as well.

**Three-dimensional**    Having length, width, and height or depth; not contained on a single plane.

**Transformation**    A general term that includes reflections, rotations, translations, and dilations; a function that carries a figure onto an image that is congruent or similar to the pre-image.

**Transitive property**    If $a = b$ and $b = c$, then $a = c$.

**Translation**    A transformation that shifts a figure horizontally and/or vertically while maintaining side lengths and angle sizes.

**Transversal**    A line that intersects two or more other lines, each in a different point.

**Trapezoid**    A quadrilateral with one pair of parallel sides.

**Triangle**    A three-sided polygon.

**Triangle Inequality theorem** In any triangle, the length of any side is less than the sum of the lengths of the other two sides but more than their difference.

**Triangle Sum theorem** The measures of the three angles of a triangle total 180°.

**Trigonometry** Literally "triangle measurement"; a branch of mathematics rooted in the ratios of sides of right triangles and dedicated to finding the unknown sides or angles of triangles.

**Truth table** A method of organizing all possible truth values for statements and gradually building to more complex statements.

**Truth value** The label, true or false, assigned to a statement to indicate whether the statement is correct.

**Two-dimensional** Having length and width, but no height or depth; lying on the plane.

**Undefined terms** The words point, line, and plane which are not formally defined, but may be described, that serve as the building block of a vocabulary for geometry.

**Valid argument** A set of statements that logically form a tautology and can be used to prove the truth of a statement.

**Vertex** In an angle, the common endpoint of the two rays that form the sides; in a polygon, a point at which two sides meet; in a polyhedron, the point at which three edges meet.

**Vertex angle** In an isosceles triangle, the angle formed by the two congruent sides.

**Vertex of parabola** The turning point of the cup-shaped graph; the highest or lowest point on the graph.

**Vertical angles** A pair of nonadjacent angles formed when two lines cross that have the point of intersection as their common vertex but do not share a side.

**Volume** The amount of space occupied by a three-dimensional figure; sometimes thought of as the amount the figure will hold, although that is technically the capacity.

***x*-axis** A horizontal number line with positive numbers to the right of zero and negative numbers to the left that is used, along with the *y*-axis, to establish a coordinate system on the plane.

***x*-coordinate** The first number in the ordered pair that locates any point in the Cartesian plane by describing how far left or right of the *y*-axis the point lies.

***y*-axis** A vertical number line with positive numbers above and negative numbers below the origin, which works with the *x*-axis to establish a coordinate system on the plane.

***y*-coordinate** The second number in the ordered pair that locates a point on the Cartesian plane, which describes how far above or below the *x*-axis the point lies.

***y*-intercept** The point(s) at which a graph crosses the *y*-axis.

# Answer Key

## Pretest

**1.** Lines, rays, or segments that meet to form a right angle are perpendicular.

**2.**

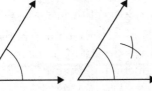

**3.** To construct a line parallel to a given line, first draw a line that crosses the given line and passes through the given point. This creates an angle at the given line, which will be copied with its vertex at the given point. It will create a congruent, corresponding angle, which will guarantee that the new line is parallel to the given line.

**4.**

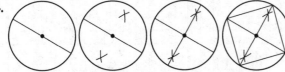

**5.** If *p*: John plays football is a true statement, and *q*: Elizabeth is a nurse is a false statement, then *p*∧*q* is false. In order for a conjunction to be true, both statements must be true.

**6.** If you score well on tests, then you study daily.

**7.** Don is trying to use syllogism, but these are not conditional statements. You could attempt to write equivalent conditionals. If you have

crackers, then you have more than nothing. If you have ice cream, then that is better than anything else. But these have the form *p*→*q* and *r*→*s* and there is no linkage for the syllogism.

**8.** If ∠*A* is supplementary to ∠*B*, then m∠*A* + m∠*B* = 180°. If ∠*B* is supplementary to ∠*C*, then m∠*B* + m∠*C* = 180°. If two expressions are both equal to the same quantity, then they are equal to each other, so m∠*A* + m∠*B* = m∠*B* + m∠*C*. If you subtract the same quantity from both sides of an equation, the new equation is still true. So if you subtract m∠*B* from both sides of the equation above, then m∠*A* = m∠*C*. If the measures of two angles are equal, the angles are congruent, so ∠*A* ≅ ∠*C*.

**9.**

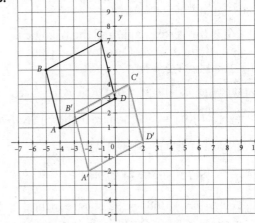

**10.** Let the *y*-axis be the line of reflection and the origin be the point on the line. Let the pre-image be the triangle with vertices $A(0,0)$, $B(3, 4)$, and $C(7, 0)$. A reflection across the *y*-axis creates a mirror image with $A'$ onto of $A$, $C'$ farthest to the left and the image triangle sitting in Quadrant II. A rotation about the origin creates an image in Quadrant III, upside down.

**11.** The orientation of the image as compared to the pre-image signals a rotation and a translation. One possible combination: rotate 90° clockwise about the origin [$A(-5, -2) \rightarrow A(-2, 5)$, $B(-5, 1) \rightarrow (1, 5)$, $C(1, 1) \rightarrow (1, -1)$, $D(1, -2) \rightarrow (-2, -1)$] then a translation 6 units to the right [$A'(4, 5)$, $B'(7, 5)$, $C'(7, -1)$, $D'(4, -1)$].

**12.** You need either $\overline{SR} \cong \overline{YZ}$ or $\overline{ST} \cong \overline{YX}$ to show $\triangle RST \cong \triangle ZYX$.

**13.** If you have the correct corresponding parts congruent to support AAS (two angles and a side not included between them), you could use the Third Angle theorem to show that the remaining pair of angles are congruent, and claim ASA (two sides and the included angle).

**14.** When a rigid transformation carries one triangle onto another triangle, the parts that align are corresponding parts.

**15.** First prove $\triangle BFC \cong \triangle DGC$ by ASA, then by CPCTC, $\overline{BF} \cong \overline{DG}$. With that, $\triangle ABF \cong \triangle EDG$ by SSS.

**16.** A scalene triangle has three sides that are different lengths.

**17.** A median of a triangle is a line segment that connects a vertex to the midpoint of the opposite side.

**18.** 233°   The three angles of a triangle always add to 180°, but $42° + 84° + 107° = 233°$.

**19.** $\angle 71°$   $B$ is the vertex angle and the triangle is isosceles, so $m\angle A = m\angle C$, so $m\angle A = (180° - 38°) \div 2 = 142° \div 2 = 71°$.

**20.** 92 square inches The area of $\triangle RST$ is

$\frac{1}{2}bh = \frac{1}{2}(RT)(h) = 368$ square inches. The area

of $\triangle MSN$ is $\frac{1}{2}(MN)\left(\frac{1}{2}h\right)$ and $MN = \frac{1}{2}RT$,

So $A = \frac{1}{2}\left(\frac{1}{2}RT\right)\left(\frac{1}{2}h\right) = \frac{1}{8}(RT)(h) = $

$\frac{1}{4}\left(\frac{1}{2}(RT)(h)\right) = \frac{1}{4}(368) = 92$ square inches.

**21.** 6 cm If $P$ is the centroid, then $AP:PM = 2:1$. $AP = 12$, so $PM = 6$ cm.

**22.** A parallelogram has two pairs of parallel sides, but a trapezoid has one pair of parallel sides.

**23.** If $ABCD$ is a parallelogram, then consecutive interior angles are supplementary. That means that $\angle A$ and $\angle B$ are supplementary and that $\angle B$ and $\angle C$ are supplementary. $m\angle A + m\angle B = 180°$ and $m\angle B + m\angle C = 180°$, so $m\angle A + m\angle B = m\angle B + m\angle C$. Subtract $m\angle B$ from both sides and $m\angle A = m\angle C$.

**24.** If the diagonals measure 26 cm and 38 cm, the area of the rhombus is $A = \frac{1}{2}d_1d_2 = \frac{1}{2}(26)(38) = 494$ square centimeters. Because the diagonals of a rhombus are perpendicular bisectors of each other, they divide the rhombus into 4 right triangles. Each right triangle has legs that are $\frac{1}{2}d_1$ and $\frac{1}{2}d_2$ so the area of the rhombus is $4\left(\frac{1}{2}\right)\left(\frac{1}{2}d_1\right)\left(\frac{1}{2}d_2\right) = 4\left(\frac{1}{8}\right)d_1d_2 = \frac{1}{2}d_1d_2$.

**25.** A polygon is regular if all its sides are congruent and all its interior angles are congruent.

**26.** 187.1 square centimeters   The area of the hexagon is $A = \frac{1}{2}P = \frac{1}{2}\left(6\sqrt{3}\right)(36) = 108\sqrt{3} \approx 187.1$ square centimeters.

**27.** The center of a dilation is a fixed point in the place from which rays project to define the enlargement or reduction of a figure.

**28.**

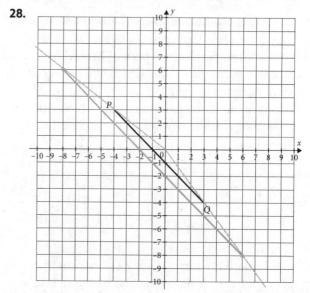

**29.** Two polygons are similar if one is a dilation of the other, in which case all corresponding angles will be congruent and all corresponding sides will be in proportion.

**30.** The lengths of corresponding segments in similar figures will always be in proportion.

**31.** In $\triangle ABC$, $m\angle C = 180° - (47° + 51°) = 82°$. In $\triangle RST$, $m\angle S = 180° - (51° + 82°) = 47°$. Because $\angle A \cong \angle S$, $\angle B \cong \angle R$, and $\angle C \cong \angle T$, $\triangle ABC \cong \triangle SRT$.

**32.** 30 cm $\dfrac{XY}{MN} = \dfrac{YZ}{NO} = \dfrac{XZ}{MO}$ becomes $\dfrac{14}{21} = \dfrac{20}{x}$. Solve to find $14x = 420$ and $x = 30$ so $MO = 30$ cm.

**33.** $WT = 12$ $\triangle VSW \sim \triangle RST$, so $\dfrac{VW}{RT} = \dfrac{SW}{ST}$.

Substitute the known lengths and solve $\dfrac{2}{10} = \dfrac{SW}{15}$ to find $SW = \dfrac{2 \cdot 15}{10} = 3$. The question asks for $WT = ST - SW = 15 - 3 = 12$.

**34.** $h = 4.8$ cm Using the known legs as base and altitude, the area is $A = \frac{1}{2}bh = \frac{1}{2}(6)(8) = 24$ square centimeters. The area is the same when calculated using the hypotenuse as the base and the altitude drawn to the hypotenuse, so $A = \frac{1}{2}bh = \frac{1}{2}(10)(h) = 24$. Simplify and solve to find $h = \dfrac{24}{5} = 4.8$ cm.

**35.** 8.7 inches $a^2 + b^2 = c^2$ becomes

$a^2 + (6.2)^2 = (10.7)^2$ or $a^2 + 38.44 = 114.49$.

Then $a^2 = 114.49 - 38.44 = 76.05$. The shorter leg, $a \approx 8.7$ inches.

**36.** $\triangle DEF$ is acute Take the two shorter sides as $a$ and $b$. $a^2 + b^2 = 4^2 + 6^2 = 16 + 36 = 52$. The longest side $c^2 = 7^2 = 49$. Because $a^2 + b^2 > c^2$, $\triangle DEF$ is acute.

**37.** 9 inches or 16 inches The altitude to the hypotenuse is the geometric mean between the sections of the hypotenuse. Let $AD = x$ and $DC = 25 - x$. Solve $\dfrac{x}{12} = \dfrac{12}{25 - x}$ by

cross-multiplying. $25x - x^2 = 144$ becomes $x^2 - 25x + 144 = 0$ and can be factored as $(x - 16)(x - 9) = 0$. This produces two solutions, $x = 16$ or $x = 9$. If $AD$ is 16, $DC$ is 9, and if $AD$ is 9, $DC$ is 16. So $AD$ could be either 9 inches or 16 inches.

**38.** All right triangles have a right angle and all right angles are congruent. If both triangles also have a 22° angle, the triangles are similar by AA.

**39.** Use the Pythagorean theorem to find that the length of the hypotenuse is 13. $\sin\angle A = \dfrac{12}{13}$,

$\cos\angle A = \dfrac{5}{13}$, and $\tan\angle A = \dfrac{12}{5}$.

**40.** $\sin\angle A = \dfrac{opposite}{hypotenuse} = \dfrac{BC}{AC} = \dfrac{12}{13}$.

$\cos\angle C = \dfrac{adjacent}{hypotenuse} = \dfrac{BC}{AC} = \dfrac{12}{13}$.

The side opposite $\angle A$ is adjacent to $\angle C$.

**41.** 15.4 cm $\sin\angle R = \dfrac{ST}{RT} = \dfrac{x}{24}$. Solve to find

$x = 24\sin(40°) \approx 15.4$ cm.

**42.** $m\angle A \approx 53.1°$ $\sin\angle A = \dfrac{BC}{AC} = \dfrac{28}{35} = 0.8$ so

$m\angle A = \sin^{-1}(0.8) \approx 53.1°$. (Or use

$\cos\angle A = \dfrac{21}{35} = 0.6$ or $\tan\angle A = \dfrac{28}{21} = \dfrac{4}{3} = 1.\overline{3}$.)

**43.** 7.1 inches Use the Law of Sines. $\dfrac{XZ}{\sin\angle Y} = \dfrac{XY}{\sin\angle Z}$

becomes $\dfrac{8}{\sin(98°)} = \dfrac{XY}{\sin(61°)}$. Cross-multiply

and solve to get $XY = \dfrac{8\sin(61°)}{\sin(98°)} \approx$

$\dfrac{8(0.8746)}{0.9903} \approx 7.1$ inches.

**44.** Use the given coordinates to find the lengths of the sides of $\triangle PQR$. $\overline{PQ}$ is horizontal and its length is $PQ = |7 - 1| = 6$. Find $QR$ and $PR$ with the distance formula.
$QR = \sqrt{(7 - 4)^2 + (1 - 5)^2} = \sqrt{9 + 16} = 5$.

$PR = \sqrt{(1 - 4)^2 + (1 - 5)^2} = \sqrt{9 + 16} = 5$.

$\triangle PQR$ is isosceles with $\overline{PR} \cong \overline{QR}$.

**45.** $y = \frac{3}{5}x - \frac{8}{5}$ First, find the midpoint of $\overline{DE}$: $m = \dfrac{y_2 - y_1}{x_2 - x_1} = \dfrac{-3 - 7}{9 - 3} = \dfrac{-10}{6} = -\dfrac{5}{3}$. To

Next find the slope of $\overline{DE}$:

$$m = \frac{y_2 - y_1}{x_2 - x_1} = \frac{-3 - 7}{9 - 3} = \frac{-10}{6} = -\frac{5}{3}. \text{ To}$$

produce a line perpendicular to $\overline{DE}$, use a slope that is the negative reciprocal of $-\frac{5}{3}$, so $m = \frac{3}{5}$.

The perpendicular bisector of $\overline{DE}$ is a line with slope $m = \frac{3}{5}$ that passes through $(6, 2)$.

$y - 2 = \frac{3}{5}(x - 6)$ or $y = \frac{3}{5}x - \frac{8}{5}$.

**46.** $y = (x + 2)^2 - 3$    A quick sketch will show that the parabola described opens up. The form of the equation of the parabola is $y - k = a(x - h)^2$. Substitute the given vertex for $h$ and $k$. $y + 3 = a(x + 2)^2$. To find the value of $a$, substitute the given point for $x$ and $y$, the solve. $1 + 3 = a(0 + 2)^2$ becomes $4 = 4a$ so $a = 1$. The equation is $y + 3 = (x + 2)^2$ or $y = (x + 2)^2 - 3$.

**47.** A triangle inscribed in a circle has three inscribed angles, each equal to half its intercepted arc. In order for it to be a right triangle, the 90° angle must have an intercepted arc that is a semicircle. The side opposite the right angle, the hypotenuse, connects the endpoints of that semicircle and so is a diameter.

**48.** 11.45 square inches    If $\angle LMN$ is inscribed in circle $O$, and measures 41°, its intercepted arc, $\overset{\frown}{LN}$, has a measure of 82°. The radius of the circle is 4, so the length of $\overset{\frown}{LN}$ is $\frac{82}{360} \cdot 2\pi \cdot 4 = \frac{82\pi}{45} \approx 5.7$ inches. The area of the sector defined by $\angle LON$ is $\frac{82}{360} \cdot \pi \cdot 4^2 = \frac{164\pi}{45} \approx 11.45$ square inches.

**49.** 5.5 grams    The volume of the cone is

$$V = \frac{1}{3}\pi r^2 h = \frac{1}{3}\pi(0.5)^2(2) = \frac{\pi}{6} \approx 0.524$$

cubic centimeters. Multiply that by 10.49 grams per cubic centimeter, the density of silver, to get $(0.524)(10.49) = 5.49676$ or approximately 5.5 grams.

**50.** 141.4 cubic inches    Rotating a rectangle about one of its sides will sweep out a cylinder. The height of the cylinder is the length of the side about which the rectangle is rotated, and the radius is the length of the other side. In this case, $r = 3$ inches and $h = 5$ inches, so the volume is $V = \pi r^2 h = \pi(3^2)(5) = 45\pi \approx 141.4$ cubic inches.

---

# CHAPTER 1

# Definitions and Construction Strategies

## EXERCISE 1.1

**1.** A line has infinite length. It continues without end in both directions. A line segment is a portion of a line between two endpoints.

**2.** A ray has one endpoint, unlike a segment, which has two. The ray continues in one direction but not two.

**3.** $\overrightarrow{TR}$. The name of a ray shows the endpoint first and then a point through which the ray passes. This ray starts at $T$ and passes through $R$. $\overrightarrow{RT}$ would be a ray starting at $R$ and passing through $T$.

**4.** $MN = 3 - (-1) = 4$.

**5.** $P$ will be $-5$    $P$ must be 4 units from $M$, so the $MP$ will also be 4 units, but must be on the opposite side of $M$ from $N$. $P$ will be $-5$.

**6.** A circle is a set of points at a fixed distance from the center. The center cannot be a fixed distance from itself, so the center cannot be a point of the circle.

**7.** 63 cm    If $\angle AOB$ measures 90°, it is $\frac{90}{360}$ or $\frac{1}{4}$ of the degrees in the circle. Therefore, $\overset{\frown}{AB}$ will be $\frac{1}{4}$ of the circumference $\frac{1}{4} \cdot 252 = 63$ cm.

## EXERCISE 1.2

**1.** In construction, the compass is the measuring tool.

**2.** Draw a line. Choose a point to be the image of *P*. Measure *PR* with your compass and scribe onto the image.

**3.** $\overline{QR}$ and $\overline{Q'R'}$ must be drawn with the same radius so that when the opening of the original angle is measured, it can be transferred to the same place on the copy.

**4.** Draw a ray to be the image of $\overrightarrow{YX}$. Scribe an arc on ∠*XYZ* and an arc of equal size on the image. Measure ∠*XYZ* with your compass and scribe onto the copy. Draw the second side from *Y'* through the intersection of the arcs.

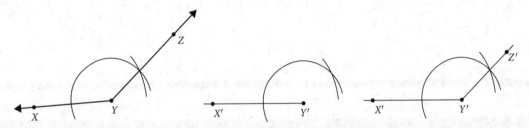

**5.** It is possible to relabel *Z* as *Y'* and make it the vertex of the copy, with $\overline{ZY}$ relabeled as $\overline{Y'Y}$ forming one side of the copy angle. (This would likely be confusing, however.)

**6.** Make a copy of $\overline{PQ}$. Copy ∠*ABC* at *P'*. Use a compass to measure $\overline{P'Q'}$, and transfer to the new side. Draw the third side.

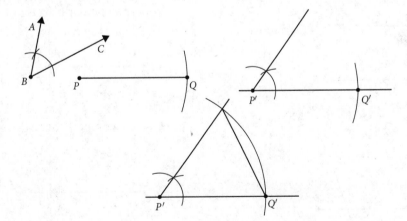

## EXERCISE 1.3

**1.** Because the arcs that intersect to locate point $P$ are drawn with the same radius, one centered at $A$ and one centered at $B$, the distances $PA = PB$.

**2.**

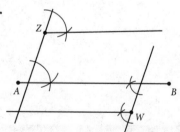

**3.**

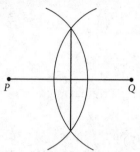

**4.** $\overline{MN}$ is parallel to the third side and half as long.

**5.**

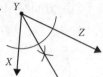

**6.**

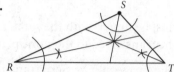

**7.** The three angle bisectors, if drawn carefully, all intersect at a single point.

## EXERCISE 1.4

**1.** Yes, $\angle RQA$ could be used instead of $\angle RQS$, as long as it was copied with its vertex at $T$ and the copy opened in the same direction as $\angle RQA$.

**2.**

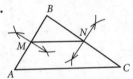

**3.**

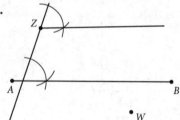

Yes, two lines which are both parallel to the same line are parallel to each other.

**4.** We want the given point to be the midpoint of some segment, so that we can perform the construction that creates a perpendicular bisector and be certain it will pass through the given point. The length of the segment we bisect is unimportant, as long as half of it is on one side and half on the other side of the given point.

**5.** Find $O$ so that $OP = PN$. Construct perpendicular bisector of $\overline{ON}$.

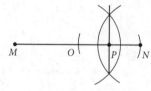

**6.** Use a compass to measure $PN$ and transfer to $PQ$.

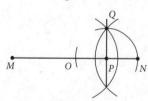

**7.** The triangle is a scalene right triangle.

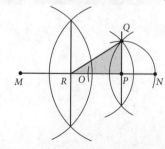

**EXERCISE 1.5**

**1.** From *T* and *R*, scribe arcs with radius *TR*. Draw from *T* and *R* to the intersection of the arcs.

**2.** Extend $\overline{XW}$ and erect perpendiculars at *X* and *W*. Measure *XW* with a compass, and scribe on each of the perpendiculars. Connect with the fourth side.

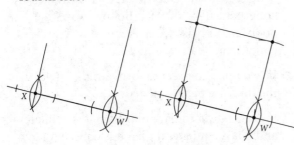

**3.** Construct the perpendicular bisector of $\overline{XW}$. The vertex between the two congruent sides will be on the perpendicular bisector. Choose any point on the bisector and connect to *X* and to *W*.

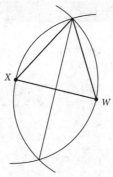

**4.** Scribe a circle with radius $\overline{TR}$. Choose any point on the circle to be the third

vertex because every radius with be congruent.

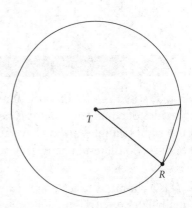

**5.** The bisector of ∠*W* intersects the opposite vertex and will bisect the opposite angle. The two triangles created by that line segment are identical.

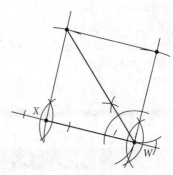

**6.** If you fold the triangle along the angle bisector, the two sides will match, so the angle bisector does bisect the opposite side.

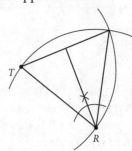

CHAPTER 2

# Logic and Proof

## EXERCISE 2.1

1. $p \land q$ is true when $\triangle ABC$ has a right angle formed by two sides of equal length. $p \land r$ is never true because right triangles always have a longest side, and because equilateral triangles have three 60° angles.

2. "$ABCD$ is a rhombus or a square" is true if $ABCD$ is a quadrilateral with four sides of equal length. Four sides of equal length guarantee it is a rhombus, which is enough to make the statement true. If $ABCD$ also has right angles, it will be a square, which will make the statement true.

3. To negate "$ABCD$ is a rhombus or a square," we must say that neither is true, both are false. The negation is "$ABCD$ is not a rhombus and not a square."

4. True   The conditional statement is "If $\triangle ABC$ is not equilateral, then it is a right triangle." $\triangle ABC$ is not equilateral is false. $\triangle ABC$ is a right triangle is false. $F \rightarrow F$ is true.

5. True   $\triangle ABC$ is equilateral is true. $\triangle ABC$ is isosceles would be called true by most people. If three sides are congruent, then two sides are congruent. $T \rightarrow T$ is true. (But be aware that some books may say isosceles means only two sides congruent. Read carefully.)

6. If two lines intersect to form a right angle, then the lines are perpendicular.

7. If a triangle is equiangular then it is equilateral and if it is equilateral then it is equiangular.

## EXERCISE 2.2

1.

| p | q | q→p | p→q |
|---|---|-----|-----|
| T | T | T | T |
| T | F | T | F |
| F | T | F | T |
| F | F | T | T |

$q \rightarrow p$ and $p \rightarrow q$ have the same truth value when both $p$ and $q$ are true or when both $p$ and $q$ are false, and opposite truth values when $p$ and $q$ have different truth values.

2. Not a tautology.

| p | q | p→q | (p→q)∧q | [(p→q)∧q] →p |
|---|---|-----|---------|--------------|
| T | T | T | T | T |
| T | F | F | F | T |
| F | T | T | T | F |
| F | F | T | F | T |

3. $\sim p \rightarrow \sim q$ agrees with $q \rightarrow p$, while $\sim q \rightarrow \sim p$ agrees with $p \rightarrow q$.

| p | q | ~p | ~q | p→q | q→p | ~p→~q | ~q→~p |
|---|---|----|----|-----|-----|-------|-------|
| T | T | F | F | T | T | T | T |
| T | F | F | T | F | T | T | F |
| F | T | T | F | T | F | F | T |
| F | F | T | T | T | T | T | T |

4. Let $p$ represent $ABCD$ is a parallelogram; $q$: The diagonals of $ABCD$ are congruent; $r$: $ABCD$ is a rectangle; $s$: $ABCD$ is a square. The statement is $(p \land q) \rightarrow (r \lor s)$.

5. $(p \land q) \rightarrow r$ is the statement "If $\triangle RST$ is a right triangle and $RS = ST$, then m$\angle R$ = m$\angle T$.

6. For three statements, you would need to look at eight possibilities; for four statements, sixteen

possibilities, and in general, for $n$ statements, $2^n$ statements.

| p | q | r |
|---|---|---|
| T | T | T |
| T | T | F |
| T | F | T |
| T | F | F |
| F | T | T |
| F | T | F |
| F | F | T |
| F | F | F |

**7.** An "or" statement is true when both $p$ and $q$ are true, but an "xor" or exclusive or is false when $p$ and $q$ are both true.

| p | q | p∨q | p x or q |
|---|---|---|---|
| T | T | T | F |
| T | F | T | T |
| F | T | T | T |
| F | F | F | F |

## EXERCISE 2.3

**15.** If a quadrilateral is a parallelogram, then it has two pairs of parallel sides.

**16.** If a quadrilateral is not a parallelogram, then its diagonals do not bisect one another.

**17.** If a parallelogram is not a rectangle, then its diagonals are not congruent.

**18.** $P$: Two lines are perpendicular; and $q$: two lines intersect to form a right angle.

**19.** $q \rightarrow \sim p$ is the converse and $p \rightarrow \sim q$ is the inverse.

**20.** The inverse of $p \rightarrow q$ is $\sim p \rightarrow \sim q$ and the converse of the inverse is $\sim q \rightarrow \sim p$.

**21.** The converse of $\sim q \rightarrow \sim p$ is $\sim p \rightarrow \sim q$ and the inverse of the converse is $p \rightarrow q$.

## EXERCISE 2.4

**1.** The form of the argument is 1) $p \rightarrow q$, 2) $q$, 3) therefore $p$. The argument is not valid. It is reasoning from the converse.

**2.** The form of the argument is 1) $p \rightarrow q$, 2) $q \rightarrow r$, 3) $p$, 4) therefore $r$. The argument is valid. 1) and 2) tell you that $p \rightarrow r$, and that with 3) allows you to draw the conclusion in 4).

**3.** The form of the argument is 1) $p \rightarrow q$, 2) $r \rightarrow q$, 3), therefore $p \rightarrow r$. The argument is not valid, but the conclusion "if a quadrilateral has two pairs of opposite sides congruent, then it has two pairs of opposite angles congruent" is in fact true.

**4.**

| p | q | r | p→q | r→q | (p→q)∧(r→q) | p→r | [(p→q)∧(r→q)]→(p→r) |
|---|---|---|---|---|---|---|---|
| T | T | T | T | T | T | T | T |
| T | T | F | T | T | T | F | F |
| T | F | T | F | F | F | T | T |
| T | F | F | F | T | F | F | T |
| F | T | T | T | T | T | T | T |
| F | T | F | T | T | T | T | T |
| F | F | T | T | F | F | T | T |
| F | F | F | T | T | T | T | T |

**5.** The form of the argument is 1) $p{\rightarrow}q$, 2) $r{\rightarrow}q$, 3), therefore $p{\rightarrow}r$. The form of the argument is not valid, and the conclusion "if a quadrilateral is a parallelogram, then it is a rhombus" is not true.

**6.** An invalid pattern of reasoning may sometimes, but not always, lead to a conclusion which is true.

**7.** You got out of bed. Form the contrapositive of each conditional, and reason back.

## EXERCISE 2.5

**1.** $\angle DEC \cong \angle GEF$, $\angle DEF \cong \angle AEG$

**2.** m$\angle APB = 43°$, m$\angle BPC = 137°$, m$\angle CPD = 43°$, m$\angle DPA = 137°$

**3.** m$\angle QPC = 133°$   Given m$\angle BPC = 86°$, m$\angle APF = 86°$ (vertical angles). m$\angle APB = 180°$ $- 86° = 94°$. $\overrightarrow{PQ}$ bisects $\angle APB$, so m$\angle QPB = 47°$. m$\angle QPC = $m$\angle APB + $m$\angle BPC = 47° + 86° = 133°$

**4.** Not perpendicular   $\angle PRS$ and $\angle SRQ$ form a linear pair and so are supplementary. Therefore $x + 2x = 180°$, so $3x = 180°$, and $x = 60°$, $2x = 120°$. Neither angle is a right angle, therefore the lines are not perpendicular.

**5.** $x = 12°$   The angles are supplementary, therefore $2(3x + 1) + (9x - 2) = 180°$. $6x + 2 +$ $9x - 2 = 15x = 180°$ and $x = 12°$.

**6.** $117°, 63°$   Because the angles are supplementary, $x + x - 54 = 2x - 54 = 180°$. Then $2x = 234°$ and $x = 117°$, so $x - 54 = 63°$.

**7.** $x = 85°$   In the triangle with angles of 85° and 15°, the third angle measures 80°, as does its vertical angle. The third angle in the middle triangle is 70°, as is its vertical angle. In the right triangle $x = 180° - (70° + 25°) = 85°$.

## EXERCISE 2.6

**1.** m$\angle 1 = $m$\angle 3 = $m$\angle 5 = $m$\angle 7 = 108°$. m$\angle 2 = $m$\angle 4 = $m$\angle 6 = $m$\angle 8 = 72°$.

**2.** $\angle 1$ and $\angle 3$ are corresponding angles, so m$\angle 1 = $m$\angle 3$. Because they are linear pairs, and therefore supplementary, m$\angle 1 + $m$\angle 2 = 180°$ and m$\angle 3 + $m$\angle 6 = 180°$. Therefore m$\angle 1 + $m$\angle 2 = $m$\angle 3 + $m$\angle 6$ and subtracting m$\angle 1$ from one side and m$\angle 3$ from the other, m$\angle 2 = $m$\angle 6$.

**3.** m$\angle 1 + $m$\angle 8 = 180°$ and m$\angle 3 + $m$\angle 4 =180°$, so m$\angle 1 + $m$\angle 8 = $m$\angle 3 + $m$\angle 4$. Because m$\angle 1 = $m$\angle 3$, subtract and m$\angle 8 = $m$\angle 4$.

**4.** Because m$\angle 1 + $m$\angle 2 = 180°$ and m$\angle 1 = $m$\angle 3$, you can show by substituting m$\angle 3$ for m$\angle 1$ that $\angle 2$ and $\angle 3$ are supplementary.

**5.** Draw the new line and label an angle. Because when parallel lines are cut by a transversal, corresponding angles are congruent, $\angle 6 \cong \angle 9$. But $\angle 6 \cong \angle 8$, so

$\angle 8 \cong \angle 9$. Because corresponding angles are congruent, $\overleftrightarrow{AB} \parallel \overleftrightarrow{EF}$.

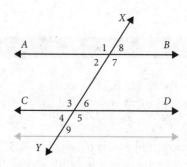

**6.** m$\angle 5 = 126°$   Because m$\angle 2 + $m$\angle 3 = 180°$, solve $(2x + 4) + (5x + 1) = 180°$ to find $x = 25°$. Then m$\angle 5 = $m$\angle 3 = 5x + 1$, m$\angle 5 = 5(25°) +$ $1 =126°$.

**7.** $\angle 1 \cong \angle 3 \cong \angle 5 \cong \angle 7 \cong \angle 9$

**8.** m$\angle 8 = 30°$   If m$\angle 9 = 150°$, then m$\angle 1 = 150°$ (corresponding angles) and its supplement m$\angle 8 = 30°$.

## EXERCISE 2.7

1. Once a right angle exists at the intersection, its vertical angle partner is also a right angle, and its supplements in the linear pairs are also right angles.

2. m∠QPR = 90°  Bisectors divide each right angle into two 45° angles. m∠QPR = m∠QPA + m∠APR = 45° + 45° = 90°.

3. Because all angles at the intersections are right angles created by the perpendicular lines, any pair of corresponding angles are congruent, and therefore $\overleftrightarrow{AB} \parallel \overleftrightarrow{CD}$. If two lines are perpendicular to the same line, they are parallel to each other.

4. Not perpendicular   m∠PTR + m∠RTQ are a linear pair, so (2x + 18) + (3x − 13) = 180°. Solve 5x + 5 = 180° to find x = 35. Then find the measure of each angle. m∠PTR = 2(35°) + 18 = 88° and m∠RTQ = 3(35°) − 13 = 92°. Neither angle is a right angle, so the lines are not perpendicular.

5. If $\overleftrightarrow{PQ} \parallel \overleftrightarrow{RS}$, when cut by transversal $\overleftrightarrow{XY}$, corresponding angles are congruent. Each of the right angles created when $\overleftrightarrow{XY}$ intersects $\overleftrightarrow{PQ}$ has a corresponding partner at the intersection of $\overleftrightarrow{XY}$ and $\overleftrightarrow{RS}$ that is also a right angle. If a line is perpendicular to one of two parallel lines, it is perpendicular to the other as well.

6. ∠APC and ∠BQE are exterior angles on the same side of the transversal, and because $\overleftrightarrow{CD} \parallel \overleftrightarrow{EF}$, they are supplementary. Solve (4x + 18) + (6x − 18) = 180° to find x = 18, and then evaluate each angle. m∠APC = 4(18°) + 18 = 90° and m∠BQE = 6(18) − 18 = 90°. The lines are intersecting at right angles and therefore are perpendicular.

7. If $\overleftrightarrow{RS} \perp \overleftrightarrow{XY}$ then ∠RAX would be 90° and 5x + 4 = 90°. Solving tells you x = 17.2°. If that is true, ∠PBY = 6(17.2) − 11 = 103.2° so $\overleftrightarrow{PQ}$ is not perpendicular to $\overleftrightarrow{XY}$ and $\overleftrightarrow{RS}$ is not parallel to $\overleftrightarrow{PQ}$. If $\overleftrightarrow{PQ} \perp \overleftrightarrow{XY}$, 6x − 11 = 90° and x = 17.2°, but m∠PBY = 92.2°. So one of the perpendiculars can exist, but not both, and if either perpendicular exists, the lines are not parallel. If the lines were parallel m∠RAX + m∠PBY would equal 180°, x = 17°, m∠RAX = 89° and m∠PBY = 91°. It is possible for any one of these conditions to exist, but only one.

---

## REVIEW 1

# Definitions, Constructions, Logic, and Proof

1. To bisect a segment is to divide it into two segments of equal length.

2. A line segment has two endpoints and a clearly defined length. A line has infinite length and continues without end in both directions. Every point on the line divides it into two rays, in opposite directions, both with infinite length. There is no unique midpoint of a line and therefore no bisector.

3. The conditional statement p→q is false when p is true and q is false.

4. Yes, ~p∧~q is the negation of p∨q.

| p | q | p∨q | ~(p∨q) | ~p | ~q | ~p∧~q |
|---|---|-----|--------|----|----|-------|
| T | T | T | F | F | F | F |
| T | F | T | F | F | T | F |
| F | T | T | F | T | F | F |
| F | F | F | T | T | T | T |

5.

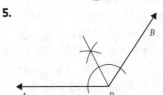

6. If △ABC is a right triangle, then $\overline{AC}$ is the longest side of △ABC or it is not.

7. m∠2 = 63°   When parallel lines are cut by a transversal, corresponding angles are congruent, so ∠9 ≅ ∠5 and ∠5 ≅ ∠7. Therefore m∠7 = 117°. ∠7 and ∠2 form a linear pair and therefore are supplementary, so m∠2 = 180° − 117° = 63°.

8.

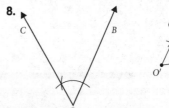

**9.** ~($p$→$q$) is the negation of $p$→$q$ and so would have the opposite truth value. The simplest way to check if ~$p$→~$q$ is equivalent to the negation is with a truth table. The table shows that the two statements are not equivalent, so ~$p$→~$q$ is not the negation of $p$→$q$.

| $p$ | $q$ | $p$→$q$ | ~$p$ | ~$q$ | ~$p$→~$q$ | ~($p$→$q$) |
|---|---|---|---|---|---|---|
| T | T | T | F | F | T | F |
| T | F | F | F | T | T | T |
| F | T | T | T | F | F | F |
| F | F | T | T | T | T | F |

To "hear" the difference, let $p$ be the statement 2 + 2 = 4 and $q$ be the statement 2 + 3 = 5. Then ~($p$→$q$) says "it is not true that 2 + 2 = 4 implies 2 + 3 = 5." ~$p$→~$q$ says "If 2 + 2 ≠ 4, then 2 + 3 ≠ 5."

**10.** If *ABCD* is not a rectangle, then *ABCD* is not a parallelogram.

**11.** Lines are perpendicular if they intersect to form a right angle.

**12.** The key piece of information missing is whether or not $\overleftrightarrow{AB} \parallel \overleftrightarrow{CD}$. If the lines are parallel, then ∠*APX* ≅ ∠*CQX* because when parallel lines are cut by a transversal, corresponding angles are congruent. If the lines are not parallel, these angles will not be congruent.

**13.** In the given statement, $p$ is "$\overline{MN}$ connects the midpoint of $\overline{AB}$ to the midpoint of $\overline{BC}$," and q is "$\overline{MN} \parallel \overline{AC}$ and $MN = \frac{1}{2} AC$." The contrapositive is ~$q$→~$p$: "If it is not true that $\overline{MN} \parallel \overline{AC}$ and $MN = \frac{1}{2} AC$, then $\overline{MN}$ does not connect the midpoint of $\overline{AB}$ to the midpoint of $\overline{BC}$," or "If $\overline{MN}$ is not parallel to $\overline{AC}$ or $MN \neq \frac{1}{2} AC$, then $\overline{MN}$ does not connect the midpoint of $\overline{AB}$ to the midpoint of $\overline{BC}$."

**14.**

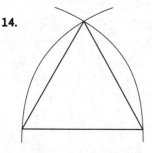

**15.**

| $p$ | $q$ | $p$→$q$ | ~($p$→$q$) | ~$q$ | $p \wedge$~$q$ | ~($p$→$q$) → $p \wedge$~$q$ |
|---|---|---|---|---|---|---|
| T | T | T | F | F | F | T |
| T | F | F | T | T | T | T |
| F | T | T | F | F | F | T |
| F | F | T | F | T | F | T |

**16.** m∠*YPW* = m∠*WPQ* = 45°   $\overline{XY} \perp \overline{MN}$ implies m∠*YPQ* = 90°. If m∠*YPW* + m∠*WPQ* = m∠*YPQ* = 90° and m∠*YPW* = m∠*WPQ* because $\overrightarrow{PW}$ bisects ∠*YPQ*, then m∠*YPW* = m∠*WPQ* = 45°.

**17.**

| $p$ | $q$ | ~$p$ | ~$q$ | $p \vee$~$q$ | ~$p \wedge q$ | ($p \vee$~$q$)→ ~$p \wedge q$) |
|---|---|---|---|---|---|---|
| T | T | F | F | T | F | F |
| T | F | F | T | T | F | F |
| F | T | T | F | F | T | T |
| F | F | T | T | T | T | F |

**18.** Two lines are parallel if they never intersect, no matter how far they are extended, or if they are always the same distance apart.

**19.** Valid. Look at the form of the argument. Let $p$ be "the parallelogram is a rhombus," let $q$ be "the diagonals of the parallelogram are perpendicular." The argument has the form [($p$→$q$)∧~$q$]→~$p$, which does not immediately match the patterns studied. But the conditional →$q$ and its contrapositive ~$q$→~$p$ are logically equivalent, so substitute: [($p$→$q$)∧~$q$]→~$p$ becomes [(~$q$→~$p$)∧~$q$]→~$p$ and the pattern of detachment is clear. If the diagonals are not perpendicular then the parallelogram is not a rhombus. The diagonals are not perpendicular; therefore, the parallelogram is not a rhombus.

**20.** The definition of the circle introduces the terms center and radius. The center, usually given as an ordered pair of coordinates, locates the circle on the plane. The radius can refer either to a line segment connecting the center to a point on the circle or to the length of such a segment. The radius determines the size of the circle.

## CHAPTER 3

# Geometry in the Coordinate Plane

### EXERCISE 3.1

**1.** $y = -\dfrac{3}{5}x + 3$   From the graph, the $y$-intercept

$b = 3$ and the slope $m = \dfrac{3 \text{ down}}{5 \text{ right}} = -\dfrac{3}{5}$, so the

equation of the line is $y = -\dfrac{3}{5}x + 3$

**2.** $m = \dfrac{-1 - 5}{5 + 4} = \dfrac{-6}{9} = -\dfrac{2}{3}$

**3.** $m = \dfrac{3 + 5}{6 - 0} = \dfrac{8}{6} = \dfrac{4}{3}$

**4.** $11x + 3y = 68$   $m = \dfrac{8 + 3}{4 - 7} = -\dfrac{11}{3}$,

then $y + 3 = -\frac{11}{3}(x - 7)$
$y + 3 = -\frac{11}{3}x + \frac{77}{3}$
$3y + 9 = -11x + 77$
$11x + 3y = 68$

**5.** $k = -5$   $m = \dfrac{k + 2}{-3 - 1} = \dfrac{k + 2}{-4} = \dfrac{3}{4}$. Then

$4(k + 2) = -4(3)$ or $k + 2 = -3$ and $k = -5$.

**6.** $y = 0$, $x = 8$, $y = 6$, and $y = 2x$.

**7.** $3x - 5y = 35$. Test the point (107, 57). $3(107) - 5(57) = 321 - 285 = 36 \neq 35$. The point is not on the line.

### EXERCISE 3.2

**1.** Perpendicular:

$2x + 3y = 15$        $3x - 2y = -14$
$3y = -2x + 15$       $-2y = -3x - 14$
$y = -\frac{2}{3}x + 5$       $y = \frac{3}{2}x + 7$

The lines are perpendicular because the slopes are negative reciprocals.

**2.** Neither parallel nor perpendicular:

$2x - 3y = 3$        $3x - 2y = -14$
$-3y = -2x + 3$       $-2y = -3x - 14$
$y = \frac{2}{3}x - 1$       $y = \frac{3}{2}x + 7$

The lines are neither parallel nor perpendicular.

**3.** Parallel:

$2x + 3y = 15$        $2x + 3y = -9$
$3y = -2x + 15$       $3y = -2x - 9$
$y = -\frac{2}{3}x + 5$       $y = -\frac{2}{3}x - 3$

The lines are parallel because they have the same slope.

**4.** $y + 5 = \frac{1}{5}(x - 10)$ or $y = \frac{1}{5}x - 7$

**5.** $y = \frac{2}{7}x$   The slope of $2x - 7y = 14$ is $\frac{2}{7}$.

Then $y - 2 = \frac{2}{7}(x - 7)$ or $y = \frac{2}{7}x$

**6.** $y = \frac{8}{5}x + 28$   Use slope of $\frac{8}{5}$.

$y - 4 = \frac{8}{5}(x + 15)$ or $y = \frac{8}{5}x + 28$

**7.** $y = \frac{3}{5}x + 1$   Slope of $5x + 3y = 0$ is $-\frac{5}{3}$.

Use slope of $\frac{3}{5}$. $y - 7 = \frac{3}{5}(x - 10)$

or $y = \frac{3}{5}x + 1$.

## EXERCISE 3.3

**1.** 5    $AB = \sqrt{(12-8)^2 + (-3+6)^2}$
   $= \sqrt{16+9} = 5$

**2.** 13    $XY = \sqrt{(0-5)^2 + (9+3)^2}$
   $= \sqrt{25+144} = 13$

**3.** 20    $RT = \sqrt{(28-16)^2 + (10-26)^2}$
   $= \sqrt{144+256} = 20$

**4.** $\triangle ABC$ is isosceles.

$AB = \sqrt{(7-3)^2 + (4-7)^2} = \sqrt{16+9} = 5,$

$BC = \sqrt{(3-7)^2 + (2-4)^2} = \sqrt{16+4} = 2\sqrt{5}$

and $AC = \sqrt{(3-3)^2 + (2-7)^2} = \sqrt{25} = 5$

**5.** $WXYZ$ is a parallelogram.

$WX\text{:} \ m = \dfrac{12-0}{5-0} = \dfrac{12}{5}, XY\text{:}\ m = \dfrac{12-12}{25-5} = 0,$

$YZ\text{:}\ m = \dfrac{12-0}{25-20} = \dfrac{12}{5}, WZ\text{:}\ m = \dfrac{0-0}{20-0} = 0.$

$\overline{WX} \parallel \overline{YZ}$ and $\overline{XY} \parallel \overline{WZ}$, so $WXYZ$ is a parallelogram.

**6.** Opposite sides are congruent.

$WX = \sqrt{(5-0)^2 + (12-0)^2}$
   $= \sqrt{25+144} = 13,$

$XY = \sqrt{(25-5)^2 + (12-12)^2}$
   $= \sqrt{400} = 20,$

$YZ = \sqrt{(25-20)^2 + (12-0)^2}$
   $= \sqrt{25+144} = 13,$

$WZ = \sqrt{(20-0)^2 + (0-0)^2}$
   $= \sqrt{400} = 20$

**7.** 240    Use $\overline{WZ}$ as the base and let the coordinates of $X$ help you see the height of the parallelogram is 12. $A = bh = (20)(12) = 240$.

## EXERCISE 3.4

**1.** $(-1, 1)$

The midpoint is $\left(\dfrac{7-9}{2}, \dfrac{-1+3}{2}\right) = (-1,1).$

**2.** $(-2.5, 7.5)$

The midpoint is $\left(\dfrac{6-11}{2}, \dfrac{13+2}{2}\right) = (-2.5, 7.5).$

**3.** $y = 13$

If $\left(\dfrac{8+x}{2}, \dfrac{-7+y}{2}\right) = (5,3)$ then $\dfrac{8+x}{2} = 5$, so

$x = 2$ and $\dfrac{-7+y}{2} = 3$, so $y = 13$.

**4.** Let $M$ be the midpoint of $\overline{AB}$ and $N$ be the midpoint of $\overline{BC}$. Then the coordinates

of $M$: $\left(\dfrac{5+7}{2}, \dfrac{3+11}{2}\right) = (6,7)$ and

$N$: $\left(\dfrac{7+15}{2}, \dfrac{11+3}{2}\right) = (11,7).$

**5.** $(10, 3)$    Let $P$ be the midpoint of $\overline{AC}$. Then

$P$: $\left(\dfrac{5+15}{2}, \dfrac{3+3}{2}\right) = (10,3).$

**6.** $\overline{MP} \parallel \overline{BC}$. $MP = \sqrt{(6-10)^2 + (7-3)^2}$
   $= \sqrt{16+16} = \sqrt{32} = 4\sqrt{2}$

and $BC = \sqrt{(7-15)^2 + (11-3)^2}$
   $= \sqrt{64+64} = \sqrt{128} = 8\sqrt{2}.$

**7.** $B$ is the point (7, 11) and $C$ is (15, 3). Divide $15 - 7 = 8$ by 4 to get 2 and divide $3 - 11 = -8$ by 4 to get $-2$. Start with $B(7, 11)$ and add (2, $-2$) to get (9, 9). Add (2, $-2$) again to get (11, 7) and again to get (13, 5).

**8.** P is ($-4$, 5) and $Q$(8, $-10$). Divide $8 - (-4) = 12$ by 3 to get 4 and $-10 - 5 = -15$ by 3 to get $-5$. Dividing points are $(-4 + 4, 5 - 5) = (0, 0)$ and $(0 + 4, 0 - 5) = (4, -5)$.

## EXERCISE 3.5

**1.** The diagonal from (0, 0) to $(a, b)$ has length $\sqrt{(a - 0)^2 + (b - 0)^2} = \sqrt{a^2 + b^2}$. The diagonal from (0, $b$) to $(a, 0)$ has length $\sqrt{(a - 0)^2 + (0 - b)^2} = \sqrt{a^2 + b^2}$. The diagonals are congruent.

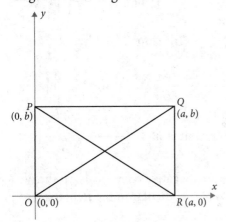

**2.** The diagonal from the origin to $(a, a)$ has a slope of $m = \dfrac{a - 0}{a - 0} = 1$. The diagonal from (0, $a$) to $(a, 0)$ has a slope of $m = \dfrac{0 - a}{a - 0} = \dfrac{-a}{a} = -1$.

The slopes are negative reciprocals, so the diagonals are perpendicular.

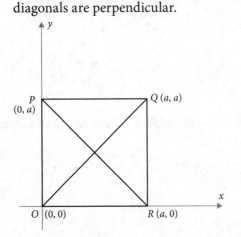

**3.** The slope of the short base is

$$m = \frac{(2b + 1) - b}{a - 0} = \frac{b + 1}{a}.$$ The slope of the long

base is $m = \dfrac{(2b + 2) - 0}{2a - 0} = \dfrac{2(b + 1)}{2a} = \dfrac{b + 1}{a}.$

The bases have the same slope and so are parallel, therefore the quadrilateral is a trapezoid.

**4.** Let $M$ and $N$ be the midpoints of the nonparallel sides. $M$: $\left(\dfrac{0 + 0}{2}, \dfrac{0 + b}{2}\right) = \left(0, \dfrac{b}{2}\right)$ and

$N$: $\left(\dfrac{a + 2a}{2}, \dfrac{(2b + 1) + (2b + 2)}{2}\right) = \left(\dfrac{3a}{2}, \dfrac{4b + 3}{2}\right).$

The slope of $\overline{MN}$ is

$$m = \frac{\left(\frac{b}{2}\right) - \left(\frac{(4b + 3)}{2}\right)}{0 - \left(\frac{3a}{2}\right)}$$

$$= \frac{\frac{-3b - 3}{2}}{\frac{-3a}{2}} = \frac{-3b - 3}{2} \cdot \frac{2}{-3a}$$

$$= \frac{-3(b + 1) \cdot 2}{2(-3a)} = \frac{b + 1}{a}.$$ The midsegment

has the same slope as the bases and therefore is parallel to the bases.

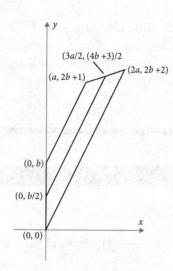

**5.** If $\overline{BD}$ bisects $\overline{AC}$., and A is $(0,0)$, then $D$ is $(a, 0)$ and C is $(2a, 0)$. $\overline{BD}$ is the perpendicular bisector of $\overline{AC}$., so $B$ is the point $(a, b)$.

$AB = \sqrt{(a-0)^2 + (b-0)^2} = \sqrt{a^2 + b^2}$ and

$BC = \sqrt{(2a-a)^2 + (0-b)^2} = \sqrt{a^2 + b^2}$.

$AB = BC$ so the triangle is isosceles.

**6.** The midpoints are $M$: $\left(\dfrac{0+a}{2}, \dfrac{0+b}{2}\right) = \left(\dfrac{a}{2}, \dfrac{b}{2}\right)$,

$N$: $\left(\dfrac{a+c}{2}, \dfrac{b}{2}\right)$, and $P$: $\left(\dfrac{c}{2}, 0\right)$

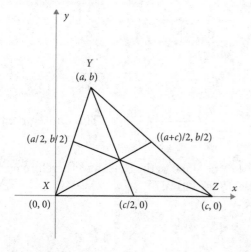

**7.** The median connects vertex $X(0,0)$ to midpoint $N\left(\dfrac{a+c}{2}, \dfrac{b}{2}\right)$. Find the increment by

$\left(\dfrac{a+c}{2} - 0\right) \div 3 = \dfrac{a+c}{6}$ and $\left(\dfrac{b}{2} - 0\right) \div 3 = \dfrac{b}{6}$.

Then begin at the origin, and the first dividing point is $\left(\dfrac{a+c}{6}, \dfrac{b}{6}\right)$ then $\left(\dfrac{a+c}{3}, \dfrac{b}{3}\right)$, and end at $N\left(\dfrac{a+c}{2}, \dfrac{b}{2}\right)$.

**8.** Place the right-angle vertex at the origin, and the acute angle vertices at $(a, 0)$ and $(0, a)$. The midpoint of the hypotenuse is $\left(\dfrac{a+0}{2}, \dfrac{0+a}{2}\right) = \left(\dfrac{a}{2}, \dfrac{a}{2}\right)$. The segment connects $(0, 0)$ to $\left(\dfrac{a}{2}, \dfrac{a}{2}\right)$. The slope of the hypotenuse is $m = \dfrac{0-a}{a-0} = \dfrac{-a}{a} = -1$, and the slope of

the new segment is $m = \dfrac{a/2 - 0}{a/2 - 0} = \dfrac{a/2}{a/2} = 1$.

The slopes are negative reciprocals, so the new segment is perpendicular to the hypotenuse.

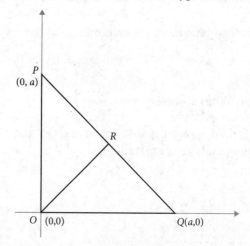

## CHAPTER 4

# Transformations

**1.** From $P$, scribe an arc to cut off $\overline{XY}$. Construct $\overline{PQ}$, the perpendicular bisector of $\overline{XY}$. Measure the distance from line $AB$ to $P$ and scribe it onto $\overline{PQ}$ to locate $P'$.

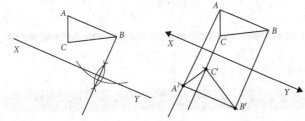

**2.** Reflect each vertex of the triangle, as above. Connect the three image points to create $\triangle A'B'C'$.

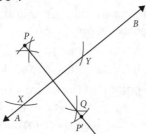

**3.** $A(4,5) \rightarrow (-4, 5)$, $B(-3, 2) \rightarrow (3,2)$, $C(2, -3) \rightarrow (-2,-3)$

**4.** $R(1,4) \rightarrow (1, -4)$, $S(3,1) \rightarrow (3, -1)$, $T(2, -4) \rightarrow (2, 4)$

**5.** $W(-1, 1) \rightarrow (1, -1)$, $X(7, 1) \rightarrow (1,7)$, $Y(7, -2) \rightarrow (-2,7)$, $Z(-1, -2) \rightarrow (-2, -1)$

**6.** $AB = \sqrt{(4+3)^2 + (5-2)^2} = \sqrt{49+9} = \sqrt{58}$,

$A'B' = \sqrt{(-4-3)^2 + (5-2)^2} = \sqrt{49+9} = \sqrt{58}$

**7.** $\triangle R'S'T'$ is oriented counterclockwise. Reflection reverses orientation.

**8.** In $WXYZ$, $\overline{WX}$ is horizontal and $\overline{XY}$ is vertical. In $W'X'Y'Z'$, $\overline{W'X'}$ is vertical and $\overline{X'Y'}$ is horizontal.

**1.** Translation is the result of two reflections. Each reflection preserves lengths and angle measures.

**2.** From $P$, scribe an arc to define a segment of $\overline{AB}$. Construct the perpendicular bisector of the segment. Mark $P'$ at an equal distance on the other side of $\overline{AB}$. $P'$ falls on $\overline{CD}$, so the second reflection over $\overline{CD}$ leaves $P'$ unchanged.

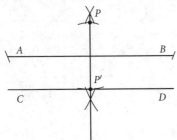

**3.** Scribe out a segment of $\overline{CD}$. Construct the perpendicular bisector. Measure from $\overline{CD}$ to $P$ and locate $P'$ at the same distance on the other side of $\overline{CD}$. Measure the distance from $P'$ to $\overline{AB}$ and locate $P''$ at the same distance on the other side of $\overline{AB}$. The result is very different from the result of exercise 10.

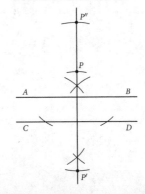

**4.** Under the translation $T_{x+5, y-2}$, $(-3, 1) \rightarrow (2, -1)$.

**5.** Under the translation $T_{x, y-4}$, $(2, 0) \rightarrow (2, -4)$.

**6.** $A(2, -3) \rightarrow (5, -5)$, $B(-1, 7) \rightarrow (2, 5)$, $C(4, 3) \rightarrow (7, 1)$

**7.** $A(-1, 1) \rightarrow (3, 0)$, $B(7, 1) \rightarrow (11, 0)$, $C(7, -2) \rightarrow (11, -3)$, $D(-1, -2) \rightarrow (3, -3)$.

**8.** Many answers are possible. Example:
$T(x, y) \rightarrow (x + 5, y - 3)$ followed by

$T(x, y) \rightarrow (x - 3, y + 4)$ followed

by $T(x, y) \rightarrow (x - 2, y - 1)$.

## EXERCISE 4.3

**1.** $R_{180°}$ is a rotation halfway around the circle centered at the origin. Whether the rotation is clockwise or counterclockwise, the endpoint is the same.

**2.** $A(2, -3) \rightarrow (-3, -2)$, $B(-1, 7) \rightarrow (7, 1)$, $C(4, 3) \rightarrow (3, -4)$

**3.** $R_{-90°}$ (a rotation of 90° clockwise)

**4.** $X'(-2, 3)$, $Y'(1, -7)$, $Z'(-4, -3)$

**5.** Reflection over the $y$-axis will only affect the $x$-coordinates, changing $x$ to $-x$, but

a 180° rotation would change both $x$- and $y$-coordinates, mapping $(x, y) \rightarrow (-x, -y)$.

**6.** $A(-1, 1) \rightarrow (-1, -1)$, $B(7, 1) \rightarrow (-1, 7)$, $C(7, -2) \rightarrow (2, 7)$, $D(-1, -2) \rightarrow (2, -1)$

**7.** Translate point $P$ in the figure 4 units left and 2 units up: $T(x, y) \rightarrow (x - 4, y + 2)$. That will put $P$ at the origin. Perform the rotation about the origin, then translate back: $T(x, y) \rightarrow (x + 4, y - 2)$

## EXERCISE 4.4

**1.** $(4, -1) \rightarrow (1, -4)$ under the rotation, then $\rightarrow (6, 0)$ under the translation.

**2.** The effect is to reflect over the $y$-axis.
$W(1, 3) \rightarrow (-1, -3) \rightarrow (-1, 3)$, $X(1, 5) \rightarrow (-1, -5) \rightarrow (-1, 5)$, $Y(5, 5) \rightarrow (-5, -5) \rightarrow (-5, 5)$, $Z(5, 3) \rightarrow (-5, -3) \rightarrow (-5, 3)$. The effect is to reflect over the $y$-axis.

**3.** The result is the same: reflection over the $y$-axis.

**4.** Over the $y$-axis, which is $x = 0$, or over any line that passes through a vertex and perpendicular to the opposite side.

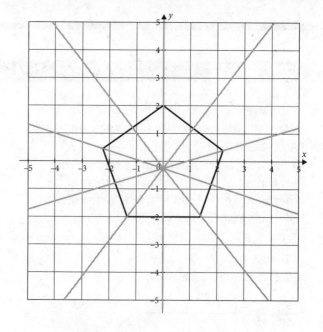

**5.** A rotation, clockwise or counterclockwise, of 60°, 120°, 180°, 240°, 300°, or 360°.

**6.** The transformations are $R_{y=x}$ followed by $T(x, y) \rightarrow (x, y - 3)$. Looking at the pre-image and image, notice B is at the top but B′ is at the bottom, which suggests reflection or rotation. Also note that the x-coordinates of the image points match the y-coordinates of the pre-image. That suggests a reflection over y = x. That reflection would map A(−1, −2)→(−2, −1), B(0, 3)→(3, 0), and C(2, 2)→(2, 2). The x-coordinates are correct, but the triangle needs to be translated down 3 units. The transformations are $R_{y=x}$ followed by $T(x, y) \rightarrow (x, y - 3)$.

**7.** *O maps onto itself when rotated through any angle. Note that results might be different if a different font had been chosen.

| | A | B | C | D | E | F | G | H | I | J | K | L | M | N | O | P | Q | R | S | T | U | V | W | X | Y | Z |
|---|---|---|---|---|---|---|---|---|---|---|---|---|---|---|---|---|---|---|---|---|---|---|---|---|---|---|
| Vertical reflection | × | | | | | | | × | × | | | | × | | × | | | | | × | × | × | × | × | × | |
| Horizontal reflection | | × | × | × | × | | | × | × | | | | | | × | | | | | | | | | × | | |
| Rotation | | | | | | | | 180° | 180° | | | | | | * | | | | 180° | | | | | 180° | | 180° |

---

# REVIEW 2

# Coordinate Geometry and Transformations

**1.**

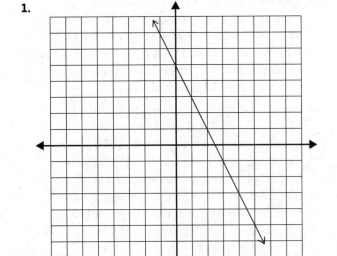

**2.** $m = \dfrac{y_2 - y_1}{x_2 - x_1} = \dfrac{5+1}{5+3} = \dfrac{6}{8} = \dfrac{3}{4}$

**3.** m = 0, b = 1  Put the equation in slope-intercept form before deciding. 2x + 3y = 2(x −1) + 5 becomes 2x + 3y = 2x − 2 + 5 or 3y = 3, which simplifies to y = 1. The slope is 0 and the y-intercept is 1.

**4.** y − 5 = −2(x + 4) or y +7 = −2(x − 2)

$m = \dfrac{y_2 - y_1}{x_2 - x_1} = \dfrac{5 + 7}{-4 - 2} = \dfrac{12}{-6} = -2$ and

then y − 5 = −2(x + 4) or y +7 = −2(x − 2) are point-slope forms of the line.

**5.** 2x − 5y = −1

$m = \dfrac{y_2 - y_1}{x_2 - x_1} = \dfrac{-1 - 1}{-3 - 2} = \dfrac{-2}{-5} = \dfrac{2}{5}$, then

$y + 1 = \frac{2}{5}(x + 3)$. Simplify and put the equation in general or standard form. $y + 1 = \frac{2}{5}x + \frac{6}{5}$ becomes 5y + 5 = 2x + 6 or 2x − 5y = −1.

**6.** Perpendicular  Put both equations in slope-intercept form. 2x − 7y = −7 becomes $y = \frac{2}{7}x + 1$ and 7x + 2y = −10 becomes $y = -\frac{7}{2}x - 5$. Because $\frac{2}{7}\left(-\frac{7}{2}\right) = -1$, the lines are perpendicular.

**7.** $y = \frac{1}{5}x + \frac{7}{5}$  5y = x + 17 has a slope of $\frac{1}{5}$, so a parallel line through (3, 2) is $y - 2 = \frac{1}{5}(x - 3)$ or $y = \frac{1}{5}x - \frac{3}{5} + 2$ The equation is $y = \frac{1}{5}x + \frac{7}{5}$.

**8.** (1, 1)  The midpoint is $\left(\dfrac{6 - 4}{2}, \dfrac{-5 + 3}{2}\right) = (1,1)$.

**9.** $y = x$   The slope of the line connecting $(-7, 3)$

to $(3, -7)$ is $m = \dfrac{y_2 - y_1}{x_2 - x_1} = \dfrac{-7 - 3}{3 + 7} = -1$,

so the slope of the perpendicular bisector will be $m = 1$. The midpoint, through which the perpendicular bisector will pass, is $M$ $\left(\dfrac{-7 + 3}{2}, \dfrac{3 - 7}{2}\right) = (-2, -2)$. The equation of the perpendicular bisector is $y + 2 = 1(x + 2)$ or $y = x$.

**10.** To show that the diagonals are congruent, use the distance formula to show they have the same length. $OQ = \sqrt{(a - 0)^2 + (b - 0)^2} = \sqrt{a^2 + b^2}$ and $PR = \sqrt{(a - 0)^2 + (0 - b)^2} = \sqrt{a^2 + b^2}$.

**11.**

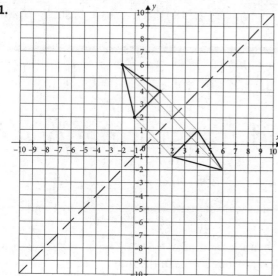

**12.**

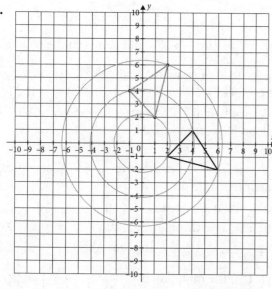

**13.** The translation moves the triangle 6 units left and 4 units down. $T(x, y) \rightarrow (x - 6, y - 4)$.

**14.** The result is not the same. If the translation is done first, each point $(a, b)$ is sent to $(a + 3, b)$. The rotation then sends to $(a + 3, b)$ to $(b, -(a + 3))$, which is equivalent to $(b, -a - 3)$. If the rotation is done first, the point $(a, b)$ is sent to $(b, -a)$ and then $(b, -a)$ is sent to $(b + 3, -a)$.

**15.**

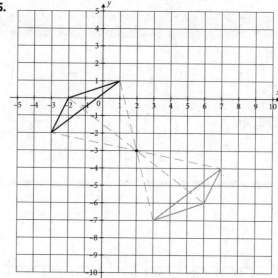

**16.** Translation 4 units right   Focus on the endpoints. $A(-1, -1)$ is reflected over $x = 3$ to $(7, -1)$, then over $x = 5$ to $(3, -1)$. $B(1, 1)$ is reflected to $(5, 1)$ and then is on the line $x = 5$, so remains at $(5, 1)$ in the second reflection. The effect is a translation 4 units right.

**17.** $A'$ is $(1, 1)$ and $B'$ is $(-1, -1)$   In this case, $A(-1, -1)$ and $B(1, 1)$ are both on the line $y = x$, and segment $\overline{AB}$ is perpendicular to $y = -x$. Reflecting over $y = -x$ sends $A(-1, -1)$ to $(1, 1)$ and $B(1, 1)$ to $(-1, -1)$. $A$ and $B$ trade places, but the change to the segment is not immediately visible. Reflecting over the line $y = x$ has no effect because $\overline{AB}$ is a segment of $y = x$. $A'$ is $(1, 1)$ and $B'$ is $(-1, -1)$.

**18.** 180°

**19.**

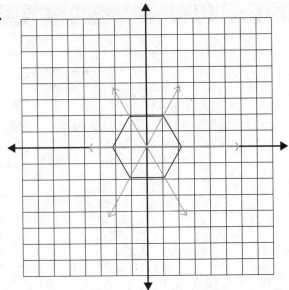

**20.** Rotations of 60°, 120°, 180°, 240°, 300°, and 360°, clockwise or counterclockwise, will carry the polygon onto itself.

# CHAPTER 5

# Triangles

## EXERCISE 5.1

**1.** $\overline{AB}$ is the base.

**2.** Vertex angle $\angle C$ is larger than either base angle, $\angle A$ or $\angle B$.

**3.** $\triangle ABC$ is an isosceles obtuse triangle.

**4.** Place the original triangle $\triangle ABC$ on the coordinate plane, with $\overline{AC}$ on the $x$-axis and point $B$ on the $y$-axis. $M$ is the point $\left(\dfrac{a}{2}, \dfrac{b}{2}\right)$ and $N$ is $\left(\dfrac{c}{2}, \dfrac{b}{2}\right)$. You can show $\overline{MN} \parallel \overline{AC}$ because both have a slope of zero, and

$$MN = \frac{c}{2} - \frac{a}{2} = \frac{c-a}{2}, \text{ which is half of } AC.$$

Each side of the small triangle is half of a side

of the larger triangle so relationships of equality or inequality will be preserved. Both triangles contain $\angle B$, and the parallel lines assure that $\angle BMN \cong \angle A$ and $\angle BNM \cong \angle C$.

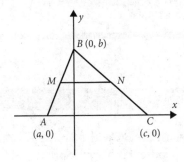

**5.** $\overline{BD}$

**6.** $\overline{BF}$

**7.** $\overline{BE}$

## EXERCISE 5.2

**1.** $m\angle R = 53°$, $m\angle S = 102°$, and $m\angle T = 25°$ The sum of the angles is 180°. $(3x - 7) + (5x + 2) + (x + 5) = 180°$ simplifies to $9x = 180°$ so $x = 20°$. Therefore, $m\angle R = 3(20) - 7 = 53°$, $m\angle S = 5(20) + 2 = 102°$, and $m\angle T = 20 + 5 = 25°$.

**2.** $\triangle RST$ is obtuse scalene.

**3.** $m\angle B = 15°$, $m\angle A = 75°$, $m\angle C = 90°$ $m\angle A + m\angle B + m\angle C = 180°$. Replace $m\angle C$ with $m\angle A + m\angle B$, and replace $m\angle A$ with $5m\angle B$. $m\angle A + m\angle B + (m\angle A + m\angle B) = 2m\angle A + 2m\angle B = 2(5m\angle B) + 2m\angle B = 12m\angle B = 180°$. Solve to find $m\angle B = 15°$. Then $m\angle A = 5(15°) = 75°$ and $m\angle C = 15° + 75° = 90°$.

**4.** $\triangle ABC$ is a right scalene triangle.

**5.** $m\angle X = 44°$, $m\angle Z = 46°$, $m\angle Y = 90°$ The acute angles of a right triangle are complementary, so $m\angle X + m\angle Z = 90°$. Therefore, $(5a - 16) + (70 - 2a) = 90°$. Simplify to $3a + 54 = 90°$ and solve to find $3a = 36°$ and $a = 12°$. $m\angle X = 5(12) - 16 = 44°$, $m\angle Z = 70 - 2(12) = 46°$ and $m\angle Y = 90°$.

**6.** In the figure below, $m\angle A = m\angle 1 + m\angle 4$ and $m\angle C = m\angle 2 + m\angle 3$.

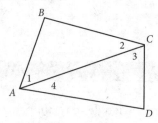

In $\triangle ABC$, $m\angle 1 + m\angle B + m\angle 2 = 180°$.
In $\triangle CDA$, $m\angle 3 + m\angle D + m\angle 4 = 180°$.
Adding the equations: $m\angle 1 + m\angle B + m\angle 2 +$

$m\angle 3 + m\angle D + m\angle 4 = 180° + 180°$. Simplify and regroup to get $(m\angle 1 + m\angle 4) + m\angle B + (m\angle 2 + m\angle 3) + m\angle D + = 360°$, or $m\angle A + m\angle B + m\angle C + m\angle D = 360°$.

**7.** In pentagon $ABCDE$, draw diagonals $\overline{CA}$ and $\overline{CE}$. Label angles as shown.

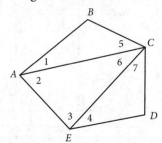

In $\triangle ABC$: $m\angle 1 + m\angle B + m\angle 5 = 180°$. In $\triangle ACE$: $m\angle 2 + m\angle 6 + m\angle 3 = 180°$. In $\triangle EDC$: $m\angle 4 + m\angle 7 + m\angle D = 180°$. Therefore, $m\angle 1 + m\angle B + m\angle 5 + m\angle 2 + m\angle 6 + m\angle 3 + m\angle 4 + m\angle 7 + m\angle D = 180° + 180° + 1A80°$. Simplify and regroup: $(m\angle 1 + m\angle 2) + m\angle B + (m\angle 5 + m\angle 6 + m\angle 7) + m\angle D + (m\angle 3 + m\angle 4) = 540°$ becomes $m\angle A + m\angle B + m\angle C + m\angle D + m\angle E = 540°$.

**8.** In any polygon with $n$ sides, if all possible diagonals are drawn from a single vertex, the polygon will be divided into $n - 2$ triangles. Interior angles of the polygon are divided in nonoverlapping ways. Each triangle has three angles that add to 180°, giving a total of $180(n - 2)°$ for the sum of the interior angles of the polygon.

## EXERCISE 5.3

**1.** Begin with $\triangle ABC$ with $\angle A = \angle C$. Construct the perpendicular bisector of $\overline{AC}$ and reflect the triangle over that line $\overline{BD}$ to create $\triangle C'B'A'$ (shown separately below). $B$ maps to $B'$ and $D$ to $D'$ because they are on the reflecting line and so remain in place. Because $D$ is the midpoint of $\overline{AC}$ and reflection preserves length, $A$ will be carried on to $C$ and $C$ to $A$. The reflection will

carry $\overline{AB}$ onto $\overline{B'C'}$ (which is the same length as $\overline{BC}$). Therefore, $\overline{AB} \cong \overline{BC}$.

**2.** $m\angle P = m\angle Q = 43°$ In $\triangle PQR$, $m\angle P + m\angle Q + m\angle R = 180°$, $m\angle P = m\angle Q$, and $m\angle R = 94°$. Substitute and solve: $2m\angle P + 94° = 180°$ means $2m\angle P = 86°$, so $m\angle P = m\angle Q = 43°$.

**3.** 106°   In $\triangle ABC$, both base angles measure 37°. $2(37°) + x = 180°$ so $x = 180° - 74° = 106°$.

**4.** $\angle R \cong \angle S \cong \angle T$   If $\triangle RST$ is equilateral, then $\overline{RS} \cong \overline{ST}$ implies $\angle T \cong \angle R$ by the Base Angle theorem. If $\overline{ST} \cong \overline{RT}$, then $\angle R \cong \angle S$ by the same theorem. Therefore, $\angle R \cong \angle S \cong \angle T$.

**5.** If $\overline{AB} \cong \overline{BC}$, then by the Base Angle theorem, $\angle BAC \cong \angle BCA$. If $\overline{BD} \cong \overline{BE}$, by the Base Angle theorem, $\angle D \cong \angle E$. $\angle BCA$ is an exterior angle of $\triangle EBC$, and $m\angle BCA = m\angle EBC + m\angle E$. $\angle BAC \cong \angle BCA$, so $m\angle BAC = m\angle EBC + m\angle E$.

**6.** Continuing from exercise 20, $\angle BAC$ is an exterior angle of $\triangle DBA$, and so $m\angle BAC = m\angle DBA + m\angle D$. $m\angle BAC = m\angle BCA$, so by

transitivity, $m\angle DBA + m\angle D = m\angle EBC + m\angle E$. Because $\angle D \cong \angle E$, substitution is possible, so that $m\angle DBA + m\angle D = m\angle EBC + m\angle D$. Subtract $m\angle D$ from both sides and $m\angle DBA = m\angle EBC$.

**7.** Suppose $\angle X$ and $\angle Y$ are the base angles. Solve $3x - 1 = 2x + 3$ to find $x = 4$. The base angles would measure $m\angle X = m\angle Y = 3(4) - 1 = 11°$ and the vertex angle $m\angle Z = 180° - 2(11°) = 158°$. That is possible. Suppose $\angle X$ and $\angle Z$ are the base angles. The Triangle Sum theorem says $2(3x - 1) + 2x + 3 = 180°$. Solve to find $x = 22.375$, and $m\angle X = m\angle Z = 66.125°$, and $m\angle Y = 47.75°$. This is also possible. Finally, if $\angle Y$ and $\angle Z$ are the base angles, solve $3x - 1 + 2(2x + 3) = 180°$, to get $x = 25°$. Because $m\angle X = 74°$, and $m\angle Y = m\angle Z = 53°$ is also possible, there is no way to determine which triangle was intended.

---

**1.** 25 cm   The centroid divides each median into sections that are $\frac{2}{3}$ and $\frac{1}{3}$ of the length of the median. If $PD = 6$ cm, then $AP = 12$ cm, and if $PE = 4$ cm, then $BP = 8$ cm. The perimeter of $\triangle AEP = AE + EP + AP = 9 + 4 + 12 = 25$ cm.

**2.** 29 cm   If $P$ is the incenter, $PQ = PR = PS = 4$ cm. $PZ = 12 - 4 = 8$ cm and $PX = 11 - 4 = 7$ cm. The perimeter of $\triangle XPZ = XZ + PZ + PX = 14 + 8 + 7 = 29$ cm.

**3.** If $\triangle ABC$ is a right triangle with hypotenuse $\overline{AC}$, legs $\overline{AB}$ and $\overline{BC}$ are both altitudes. They intersect at $B$, from which the third altitude is drawn to $\overline{AC}$. The three altitudes intersect at $B$, the vertex of the right angle, and the orthocenter.

**4.** The incenter is the intersection of the angle bisectors and the circumcenter is the intersection of the perpendicular bisectors of the sides. The angle bisector and the perpendicular bisector of the opposite side are the same line in isosceles and equilateral triangles. In order for the

incenter and circumcenter to be the same point, the triangle would need to be equilateral.

**5.** In an equilateral triangle, the bisector of an angle is also the perpendicular bisector of the opposite side. $AD = \frac{1}{2} AB$ and $EC = \frac{1}{2} BC$, $AD = EC$ because halves of equals are equal.

**6.** 50 cm   First find $x$. $CP = 2PF$ so $x + 7 = 2(x + 1)$. Simplify and solve to find $x = 5$. Then find $BP$, $CP$, and $BC$. $BP = \frac{2}{3} BE = \frac{2}{3}(4x + 1) = \frac{2}{3}(21) = 14$. $CP = x + 7 = 12$. $BC = 5x - 1 = 24$. Perimeter of $\triangle BPC = BP + CP + BC = 14 + 12 + 24 = 50$ cm.

**7.**

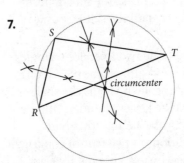

## EXERCISE 5.5

**1.** $7 < x < 43$   The third side $x$ is $25 - 18 < x < 25 + 18$ or $7 < x < 43$.

**2.** $69°$   $m\angle BCD = m\angle A + m\angle B$. $122° = 53° + x$ so $x = 69°$.

**3.** $m\angle QRT = 80°$, $m\angle^\wedge QR = 37°$   $m\angle PRQ + m\angle QRT = 180°$, so solve $7x + 2 + 6x - 4 = 180°$ to find $x = 14°$. The exterior angle $\angle QRT$ measures $6(14°) - 4 = 80°$, and $m\angle PQR = 80° - 43° = 37°$.

**4.** $\angle^\wedge C$ is an exterior angle of $\triangle ABE$, and so $m\angle AEC = m\angle BAE + m\angle B$. Because $\overline{AE}$ is an angle bisector, $m\angle BAE = m\angle EAC$. Substitute to get $m\angle AEC = m\angle EAC + m\angle B$. Subtract to get $m\angle AEC - m\angle EAC = m\angle B$.

**5.** $AD < AB < BD < CD < BC$   In $\triangle ABD$, $AD < AB < BD$. In $\triangle DBC$, $BD < CD < BC$. Therefore, $AD < AB < BD < CD < BC$.

**6.** $m\angle A > m\angle B > m\angle C > m\angle E > m\angle D$. In $\triangle ABC$, $m\angle A > m\angle B > m\angle C$. In $\triangle CDE$, $m\angle C > m\angle E > m\angle D$. Therefore, $m\angle A > m\angle B > m\angle C > m\angle E > m\angle D$.

**7.** Given that exterior angles $\angle BAD \cong \angle BCE$. By the Exterior Angle theorem, $m\angle BAD = m\angle ABC + m\angle BCA$ and $m\angle BCE = m\angle ABC + m\angle BAC$. Because $m\angle BAD = m\angle BCE$, substituting gives $m\angle ABC + m\angle BCA = m\angle ABC + m\angle BAC$, and subtracting $m\angle ABC$ from both sides leaves $m\angle BCA = m\angle BAC$. By the Base Angle Converse theorem, $AB = BC$, and $\triangle ABC$ is isosceles.

**8.** $360°$   By the Exterior Angle theorem, $m\angle 1 = m\angle BAC + m\angle BCA$, $m\angle 2 = m\angle BAC + m\angle CBA$, and $m\angle 3 = m\angle CBA + m\angle BC^\wedge$. Add to get $m\angle 1 + m\angle 2 + m\angle 3 = (m\angle BAC + m\angle BCA) + (m\angle BAC + m\angle CBA) + (m\angle CBA + m\angle BCA)$. Simplify to $m\angle 1 + m\angle 2 + m\angle 3 = 2m\angle BAC + 2m\angle BAC + 2m\angle BCA = 2(m\angle BAC + m\angle BAC + m\angle BCA)$. By the Triangle Sum theorem, $m\angle 1 + m\angle 2 + m\angle 3 = 2(m\angle BAC + m\angle BAC + m\angle BCA) = 2(180°) = 360°$.

**9.** $\overline{QR}$   The right angle, $\angle Q$, will be the largest angle, so $m\angle Q > m\angle R > m\angle P$, which means $PR > PQ > QR$. $\overline{QR}$ is the shortest side.

## EXERCISE 5.6

**1.** $DC > AD$   $AB = BC$, $BD = BD$, but included angle $\angle ABD$ measures $43°$, while included angle $\angle CBD$ measures $46°$. $DC > AD$ because the longer side is opposite the larger angle.

**2.** $m\angle PQR > m\angle RQT$   $PQ = QR = QT$ but $PR > RT$, so the angle opposite $\overline{PR}$ is larger than the angle opposite $\overline{RT}$. $\angle PQR$ is larger than $\angle RQT$.

**3.** $AB < BC$   $\overline{BD}$ is a median, so $AD = DC$, $BD = BD$ and $m\angle ADB < m\angle BDC$. The side opposite $\angle ADB$ will be shorter than the side opposite $\angle BDC$. $AB < BC$.

**4.** $m\angle ABC > \angle ADC$   The two isosceles triangles have the same base $\overline{AC}$ and can be placed on top of one another as shown. $\overline{BE}$ is the smaller altitude and $\overline{DE}$ is the larger altitude. Although the question sounds like a Hinge theorem question, the conditions of two pair of congruent sides cannot be met. Instead, use the Exterior Angle theorem. $\angle 1$ is an exterior angle of $\triangle ADB$ and $m\angle 1 = m\angle 2 + m\angle 3$, which means $m\angle 1 > m\angle 2$. $\angle 1$ is half of the vertex angle of $\triangle ABC$

and $\angle 2$ is half of the vertex angle of $\triangle ADC$. $2m\angle 1 > 2m\angle 2$, so $m\angle ABC > \angle ADC$.

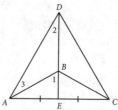

**5.** $AB = BC$ and $BD = BD$. You need to know the relative sizes of $\angle ABD$ and $\angle CBD$ to decide which of $\overline{AD}$ and $\overline{DC}$ is larger.

**6.** $\overline{XZ}$ is not the angle bisector   $WX = XY$ and $XZ = XZ$. If $\overline{XZ}$ were the bisector of $\angle WXY$, $\angle WXZ$ and $\angle YXZ$ would be congruent, and the sides opposite them, $\overline{WZ}$ and $\overline{ZY}$ would be congruent. But $WZ < ZY$, so $\overline{XZ}$ cannot be the angle bisector.

**7.** $AB < BC$   $\overline{AD} \cong \overline{CD}$ is given and $\overline{BD} \cong \overline{BD}$ because any segment is congruent to itself. $m\angle ADB = 35°$, and $m\angle CDB = 47°$, which means that $m\angle ADB < m\angle CDB$. Therefore, by the Hinge theorem, $AB < BC$.

# CHAPTER 6

# Right Triangles

## EXERCISE 6.1

**1.** Not a right triangle   $m\angle A = 17°$, $m\angle B = 74°$, $m\angle C = 180° - (17° + 74°) = 89°$. The triangle is not a right triangle.

**2.** $\overline{XY}$   $m\angle X = 47°$, $m\angle Y = 90°$, $m\angle Z = 90° - 47° = 43°$. The smallest angle is $\angle Z$, and the shortest side is $\overline{XY}$.

**3.** $m\angle A = 38°$, $m\angle B = 52°$ $m\angle A + m\angle B = (7x - 11) + (6x + 10) = 90°$. Simplify and solve to find $x = 7$. $m\angle A = 7(7) - 11 = 38°$. $m\angle B = 6(7) + 10 = 52°$.

**4.** The shortest side is $\overline{BC}$, then $\overline{AC}$, and the longest side, the hypotenuse is $\overline{AB}$.

**5.** 18.5 cm   Area $= \frac{1}{2}$ (leg 1)(leg 2) $= \frac{1}{2}$ (hypotenuse)(altitude to hypotenuse) $= \frac{1}{2}(52x) = 480$. Solve $26x = 480$ to find $x = 18.46$. The altitude to the hypotenuse is approximately 18.5 cm.

**6.** $\angle S$   $(2x + 4) + (\frac{1}{2}x) + (3x - 13) = 90$. Simplify and solve to find $5.5x = 99$, so $x = 18$. The lengths of the sides are $RS = 2(18) + 4 = 40$, $ST = \frac{1}{2}(18) = 9$, and $RT = 3(18) - 13 = 41$. The longest side is $RT$, which will be the hypotenuse so the right angle is $\angle S$.

**7.** 7 and 24   Use the given area and the altitude to the hypotenuse to solve for the length of the hypotenuse. Simplify $84 = \frac{1}{2}(6.72)h$ to $84 = 3.36h$ and solve to find $h = 25$ cm. The legs will be less than 25 cm. Area equals $\frac{1}{2}$ the product of the lengths of the legs, and area $= 84$, so the product of the lengths of the legs is $2(84) = 168$. List possible integer factors of 168 and eliminate any with numbers greater than or equal to 25. $168 = 1(168) = 2(84) = 3(56) = 4(42) = 6(28) = 7(24) = 8(21) = 12(14)$. Check the remaining three possibilities in the Pythagorean theorem. Only 7 and 24 produce a hypotenuse of 25.

## EXERCISE 6.2

**1.** 17   $a^2 + b^2 = c^2$ so $c^2 = 8^2 + 15^2 = 64 + 225 = 289$. $c = \sqrt{289} = 17$.

**2.** 96   $a^2 + b^2 = c^2$ so $b^2 = c^2 - a^2 = 100^2 - 28^2 = 10,000 - 784 = 9,216$. $b = \sqrt{9,216} = 96$.

**3.** 8.1 feet   $a^2 + b^2 = c^2$, so $c^2 = 7^2 + 4^2 = 49 + 16 = 65$. $c = \sqrt{65} \approx 8.1$ feet.

**4.** 24   $a^2 + b^2 = c^2$ becomes $10^2 + x^2 = (x + 2)^2$ or $100 + x^2 = x^2 + 4x + 4$. Simplify to $100 = 4x + 4$ and solve to find $x = 24$, which is the length of $\overline{AB}$.

**5.** 63   Let $x =$ length of leg $\overline{RS}$, $y =$ length of leg $\overline{ST}$, and $h =$ length of hypotenuse $\overline{RT}$. $A = 8x = \frac{1}{2}xy$, so $\frac{1}{2}y = 8$ and $y = 16$. $P = 2x + 18 = x + y + h$, so $2x + 18 = x + 16 + h$, or $h = x + 2$. Then $a^2 + b^2 = c^2$ becomes

$16^2 + x^2 = (x + 2)^2$, which simplifies to $256 + x^2 = x^2 + 4x + 4$, or $256 = 4x + 4$. Solve to find $4x = 252$ and $x = 63$.

**6.** a) Not a Pythagorean Triple.   $c^2 = 7^2 + 9^2 = 49 + 81 = 130$, $c = \sqrt{130} \approx 11.4$. Not a Pythagorean Triple. b) Pythagorean Triple. $a^2 = 29^2 - 21^2 = 841 - 441 = 400$, $a = \sqrt{400} = 20$. Pythagorean Triple. c) Pythagorean Triple. $b^2 = 41^2 - 9^2 = 1681 - 81 = 1600$, $b = \sqrt{1600} = 40$. Pythagorean Triple.

**7.** 18.1 feet   Ratio: 1 inch: 12 inches = 18 inches: 216 inches. Use Pythagorean theorem: $c^2 = 18^2 + 216^2 = 324 + 46,656 = 46,980$ and $c = \sqrt{46,980} \approx 216.75$. The ramp itself is approximately 216.75 inches or about 18.1 feet.

## EXERCISE 6.3

1. **Obtuse** Use the Triangle Inequality theorem to check if a triangle is possible. $5 + 8 > 11$, so these lengths can make a triangle. $5^2 + 8^2 = 25 + 64 = 89$ and $11^2 = 121$, so $a^2 + b^2 < c^2$. The triangle is obtuse.

2. **No triangle possible** Triangle inequality is not satisfied. $17 + 23 = 40 < 41$. No triangle possible.

3. **Right triangle** Triangle is possible because $39 + 52 = 91 > 65$. $39^2 + 52^2 = 1{,}521 + 2{,}704 = 4{,}225$ and $65^2 = 4{,}225$. The triangle is a right triangle.

4. **Acute** $12 + 17 > 19$, so a triangle is possible. $12^2 + 17^2 = 144 + 289 = 433$ and $19^2 = 361$. $a^2 + b^2 > c^2$. The triangle is acute.

5. **8.6. $2 < x < 8.6$** For a triangle to be possible, the third side, $x$, must meet the triangle inequality. $7 - 5 < x < 7 + 5$ or $2 < x < 12$. For the triangle to be acute, $a^2 + b^2$ must be greater than $c^2$.

$5^2 + 7^2 = 25 + 49 = 74$, so $x^2$ must be less than 74, and $x$ must be less than about 8.6. $2 < x < 8.6$.

6. **$14.4 < x < 20$** To make a triangle with sides of 8, 12, and $x$, we must have $12 - 8 < x < 12 + 8$ or $4 < x < 20$. In order for the triangle to be obtuse $a^2 + b^2 = 8^2 + 12^2 = 64 + 144 = 208$ must be less than $x^2$, so $x$ must be greater than about 14.4. $14.4 < x < 20$.

7. **$8.5 < x < 13$** In this case, $11 - 7 < x < 11 + 7$, or $4 < x < 18$. In order to have an acute triangle, either $x^2 + 7^2$ must be greater than $11^2$, or $7^2 + 11^2$ must be greater than $x^2$. Solve $x^2 + 49 > 121$ or $x^2 > 72$ to find that the first possibility would require that $x$ be greater than about 8.5, so $8.5 < x < 18$. The second possibility is that $7^2 + 11^2 > x^2$ or $x^2 < 170$. That would require $x$ to be less than about 13. That limits $x$ to $8.5 < x < 13$.

## EXERCISE 6.4

1. **$64\sqrt{2}$ cm** If the leg of a isosceles right triangle measures 64 cm, the hypotenuse measures $64\sqrt{2}$ cm.

2. **$9\sqrt{3}$ meters** If the hypotenuse of a 30°-60°-90° right triangle is 18 meters, the shorter leg is 9 meters and the longer leg is $9\sqrt{3}$ meters.

3. **$19\sqrt{2}$ inches** If the side of the square is 19 inches, the diagonal is $19\sqrt{2}$ inches.

4. **$34\sqrt{3}$ cm** If the shortest side of a 30°-60°-90° right triangle measures 34 cm, the other leg is $34\sqrt{3}$ cm.

5. **$11\sqrt{2}$ inches** If the hypotenuse of an isosceles right triangle measure 22 inches, the leg measures $\dfrac{22}{\sqrt{2}} = \dfrac{22\sqrt{2}}{2} = 11\sqrt{2}$ inches.

6. **$18\sqrt{3}$ cm** If the hypotenuse of a 30°-60°-90° right triangle is 36 cm, the longer leg measures $18\sqrt{3}$ cm.

7. **$14\sqrt{2}$ cm** If the area of the isosceles right triangle is $A = \frac{1}{2}bh = \frac{1}{2}\left(leg\right)^2 = 98$ square centimeters, then $\left(leg\right)^2 = 196$ and $leg = \sqrt{196} = 14$ cm, so the hypotenuse measures $14\sqrt{2}$ cm.

## REVIEW 3

# Triangles, Right Triangles

1. a) obtuse scalene  b) acute equilateral (equiangular)  c) right scalene  d) acute isosceles.

2. $m\angle A = 33°$, $m\angle B = 74°$, $m\angle ACB = 73°$, $m\angle BCD = 107°$. $(89 - x) + (2x + 3) + (5x - 2) =$

180° simplifies to $6x + 90 = 180°$ so $x = 15°$. $m\angle A = 33°$, $m\angle B = 74°$, $m\angle ACB = 73°$, $m\angle BCD = 107°$.

**3.** $ON < MN < MO$    $m\angle M = 58°$, $m\angle N = 63°$, and $m\angle O = 180° - (58° + 63°) = 59°$. $m\angle M < m\angle O < m\angle N$, so $ON < MN < MO$.

**4.** $m\angle R = 47°$, $m\angle T = 39°$ and $m\angle TSV = 86°$ $m\angle R + m\angle T = m\angle TSV$ becomes $(4x - 5) + 3x = 6x + 8$. Solve to find $x = 13$. $m\angle R = 4(13) - 5 = 47°$, $m\angle T = 3(13) = 39°$ and $m\angle TSV = 6(13) + 8 = 86°$.

**5.** $m\angle A = m\angle B = 70°$  If $\overline{AB} \cong \overline{BC}$, then $m\angle A = m\angle B$ by the Base Angle theorem. Solve for $x$ using either $2x + 10 = 3x - 20$ or $(2x + 10) + (3x - 20) + (x + 10) = 180$, to find $x = 30°$. $m\angle A = m\angle B = 70°$.

**6.** 73 cm.   If $XP = 28$, then $PA = 14$. If $PB = 16$, then $YP = 32$. If $YZ = 54$, then $AY = 27$. The perimeter of $\triangle PAY$ is $PA + AY + YP = 14 + 27 + 32 = 73$ cm.

**7.** By the Hinge theorem, $m\angle AOB < m\angle COD$.

**8.** $2 < x < 8$. The length of the third side is between 2 and 8 feet.

**9.** If $\angle ZXY \cong \angle ZYX$, then $\overline{XZ} \cong \overline{ZY}$ by the Base Angle Converse theorem. $\angle XZW \cong \angle YZW$ is given and $\overline{WZ} \cong \overline{WZ}$ by the reflexive postulate. $\triangle XZW \cong \triangle YZW$ by SAS and $\overline{XW} \cong \overline{YW}$ by CPCTC. $\triangle XWY$ is isosceles because it has two congruent sides.

**10.** $\triangle ACE$ is an isosceles triangle with $\overline{AC} \cong \overline{CE}$ and $\triangle BDG$ is an isosceles triangle with $\overline{BG} \cong \overline{GD}$ (given). In an isosceles triangle, base angles are congruent; therefore, $\angle HAB \cong \angle FED$ and $\angle HBD \cong \angle FDB$. Because $AE \parallel BD$, alternate interior angles are congruent, so $\angle HBD \cong \angle BHA$ and $\angle FDB \cong \angle DFE$. By the Triangle Sum theorem, $m\angle HAB + m\angle BHA + m\angle HBA = 180°$ and $m\angle FED + m\angle DFE + m\angle FDE = 180°$, so by transitivity, $m\angle HAB + m\angle BHA + m\angle HBA = m\angle FED + m\angle DFE + m\angle FDE$. Subtracting equal measures from each side gives $m\angle BHA + m\angle HBA = m\angle DFE + m\angle FDE$ and $m\angle HBA = m\angle FDE$. Because their measures are equal, $\angle HBA \cong \angle FDE$.

**11.** 81.2 inches   $a^2 + b^2 = c^2$ becomes $c^2 = 36^2 + 77^2 = 1,296 + 5,299 = 6,595$, so $c = \sqrt{6,595} \approx 81.2$ inches.

**12.** 182 cm   The legs of the right triangle are 13 cm and $h - 1$ cm, and the hypotenuse is $h$ cm. $a^2 + b^2 = c^2$ becomes $13^2 + (h - 1)^2 = h^2$ or $169 + h^2 - 2h + 1 = h^2$. Simplify and solve to find $2h = 170$ and $h = 85$. The legs are 13 cm and 84 cm. The perimeter is $85 + 84 + 13 = 182$ cm.

**13.** 7 meters   $(x - 8)^2 + (x + 9)^2 = 25^2$ becomes $x^2 - 16x + 64 + x^2 + 18x + 81 = 625$ and simplifies to $2x^2 + 2x + 145 = 625$ or $x^2 + x - 240 = 0$. Solving $(x - 15)(x + 16) = 0$ gives a solution of $-16$, which is rejected and a solution of $x = 15$. The shorter leg is $15 - 8 = 7$ meters.

**14.** Obtuse   Using the two shorter sides, $a^2 + b^2 = 8^2 + 11^2 = 64 + 121 = 185$, but $c^2 = 14^2 = 196$. Because $a^2 + b^2 < c^2$, the triangle is obtuse.

**15.** Acute   Assuming that the longest side is 2x, that there is a second side that is 2x and the shortest side is x, $a^2 + b^2 = x^2 + (2x)^2 = 5x^2$ and $c^2 = (2x)^2 = 4x^2$. $a^2 + b^2 > c^2$, so the triangle is acute.

**16.** Right   $a^2 + b^2 = x^2 + x^2 = 2x^2$ and $c^2 = \left(x\sqrt{2}\right)^2 = 2x^2$. $a^2 + b^2 = c^2$, which indicates the triangle is a right triangle.

**17.** 64.5 meters   $a^2 + b^2 = c^2$ becomes $a^2 + 100^2 = 119^2$, so $a^2 = 14,161 - 10,000 = 4,161$. Then $a = \sqrt{4,161} \approx 64.5$ meters.

**18.** Obtuse   Using the full triangle with sides of 68, 119, and 153 meters. Compare $a^2 + b^2 = 68^2 + 119^2 = 18,785$ with $c^2 = 153^2 = 23,409$. $a^2 + b^2 < c^2$, so the triangle is obtuse, not right.

**19.** $x = 10.125$   $BD$ is the geometric mean between $AD$ and $DC$, so $\dfrac{8}{9} = \dfrac{9}{x}$ and $8x = 81$; therefore, $x = 10.125$.

**20.** 40 cm   $ST$ is the geometric mean between UT and $RT$, so $\dfrac{21}{29} = \dfrac{29}{x}$ and $21x = 841$. Therefore, $x \approx 40$, so $RT$ is 40 cm.

## CHAPTER 7

# Congruence

**1.** $\angle C \cong \angle D$, $\angle A \cong \angle O$, $\angle T \cong \angle G$, $\overline{CA} \cong \overline{DO}$, $\overline{AT} \cong \overline{OG}$, $\overline{CT} \cong \overline{DG}$

**2.** $\triangle RST \cong \triangle ABC$ is not correct. $R$ corresponds to $C$, not to $A$. $T$ corresponds to $A$, not $C$. Correct statement is either $\triangle RST \cong \triangle CBA$ or $\triangle TSR \cong \triangle ABC$.

**3.** Yes, $\triangle RST \cong \triangle GHI$.

**4.** The transformation is a reflection over the line $y = x$. $\square ABCD \cong \square WZYX$.

**5.** $\angle A \cong \angle P$, $\angle B \cong \angle Q$, $\angle C \cong \angle R$, $\angle D \cong \angle S$, $\overline{AB} \cong \overline{PQ}$, $\overline{BC} \cong \overline{QR}$, $\overline{CD} \cong \overline{RS}$, $\overline{AD} \cong \overline{PS}$

**6.** Any triangle is congruent to itself because there are rigid transformations that map it onto itself, for example, a rotation of 360° about its center. Each angle is congruent to itself and each side is congruent to itself. If the triangle is $\triangle ABC$, the congruence statement is $\triangle ABC \cong \triangle ABC$.

**7.** As seen in the previous exercise, any triangle is congruent to itself. If the isosceles triangle is $\triangle ABC$ with $\overline{AB} \cong \overline{BC}$, the two possible correspondences are $\triangle ABC \cong \triangle ABC$ or $\triangle ABC \cong \triangle CBA$.

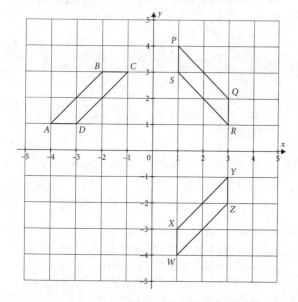

**1.** $\triangle ABC$ is given with $\overline{AC} \cong \overline{BC}$. It is given that $M$ is the midpoint of $\overline{AB}$. By the definition of midpoint, $\overline{AM} \cong \overline{MB}$. $\overline{CM} \cong \overline{CM}$ because every segment is congruent to itself. $\triangle ACM \cong \triangle BCM$ by SSS.

**2.** We cannot be certain if the triangles are congruent. Suppose $\triangle ABC$ is isosceles with $\overline{AC} \cong \overline{BC}$, and base $\overline{AC}$, and suppose $\triangle XYZ$ is isosceles with $\overline{XY} \cong \overline{YZ}$, and base $\overline{XZ}$. Assume $\overline{AC} \cong \overline{XZ}$. This meets Chris's description, but it is possible that $AC = BC = 12$ inches and $XY = YZ = 3$ inches. This is not enough information to guarantee congruent triangles.

3. It is given that $\overline{MN}$ and $\overline{PQ}$ bisect each other, so by the definition of bisector, $\overline{MO} \cong \overline{ON}$ and $\overline{PO} \cong \overline{OQ}$. $\angle MOP \cong \angle NOQ$ because vertical angles are congruent. $\triangle MPO \cong \triangle NQO$ by SAS.

4. $\overline{AC} \cong \overline{BC}$ is given. Because $\overline{CD}$ is the bisector of $\angle ACB$, $\angle ACD \cong \angle BCD$. $\overline{CD} \cong \overline{CD}$ because every segment is congruent to itself. $\triangle ACD \cong \triangle BCD$ by SAS. In an isosceles triangle, constructing the bisector of the vertex angle or the median from the vertex angle to the base or the altitude from the vertex angle to the base will all divide the triangle into two congruent right triangles. All three jobs are done by the same segment if the triangle is isosceles.

5. If $ABCD$ is a rhombus, by the definition of a rhombus, $AB = BC = CD = DA$. If diagonal $BD$ is drawn, it creates $\triangle ABD$ and $\triangle CDB$. $AB = CD$, $BD = BD$, and $DA = BC$. $\triangle ABD \cong \triangle CDB$ by SSS. Note because all sides of a rhombus are congruent, it is also possible to prove $\triangle ABD \cong \triangle CBD$ by SSS.

6. It would still be possible to prove $\triangle ABD \cong \triangle CDB$, because opposite sides of a rectangle are congruent, but it would be necessary to take care because only one correspondence is possible.

7. It can be shown that $XM = MZ$ and $YM = YM$, but it is not possible that $\angle XMY$ is congruent to $\angle ZMY$. In fact, we can argue that $\triangle XMY$ will not be congruent to $\triangle ZMY$ because the third sides, $\overline{XY}$ and $\overline{YZ}$, are known to be unequal. Without proving the triangles congruent, there is no evidence to suggest that $\overline{YM}$ is an angle bisector.

## EXERCISE 7.3

1. $\angle RST \cong \angle RQP$ and $\overline{SR} \cong \overline{QR}$ according to the given information. Because vertical angles are congruent, $\angle SRT \cong \angle QRP$. $\triangle PQR \cong \triangle TSR$ by ASA.

2. $\angle RST \cong \angle RQP$ and $\angle STR \cong \angle RPQ$

3. According to the given information, $\angle MNQ \cong \angle QPM$ and $\angle NMQ \cong \angle PQM$. $\overline{MQ} \cong \overline{MQ}$ by the reflexive property. $\triangle NMQ \cong \triangle PQM$ by AAS.

4. The fact that $\angle BAC \cong \angle DCA$ is given. $\angle B$ and $\angle D$ are right angles because in the rectangle all interior angles are right angles, and because all right angles are congruent, $\angle B \cong \angle D$. By the reflexive property, $\overline{AC} \cong \overline{AC}$. $\triangle ABC \cong \triangle CDA$ by AAS.

5. To prove $\triangle ABC \cong \triangle CDA$ with these new circumstances, it is necessary to show one of three things: 1) $\angle BCA \cong \angle DAC$ to show ASA, 2) $\angle B \cong \angle D$ to show AAS, or 3) $\overline{AB} \cong \overline{CD}$ to show SAS. All three are possible, with some information about parallelograms, but the simplest is 3) $\overline{AB} \cong \overline{CD}$ because all four sides of the rhombus are congruent.

6. It is given that $\angle NMO \cong \angle PQO$ and that $\overline{NM} \cong \overline{PQ}$. Because vertical angles are congruent, $\angle NOM \cong \angle POQ$. $\triangle NMO \cong \triangle PQO$ by AAS.

7. The given information tells about one pair of angles and one pair of sides. By the Base Angle theorem, you know $\angle A \cong \angle C$ and $\angle R \cong \angle T$. With the Triangle Sum theorem, you can show $m\angle A + m\angle B + m\angle C = 2m\angle A + m\angle B = 180°$ and $m\angle R + m\angle S + m\angle T = 2m\angle R + m\angle S = 180°$. With the transitive property, $2m\angle A + m\angle B = 2m\angle R + m\angle S$. $m\angle B = m\angle S$ (given) so by subtracting, $2m\angle A = 2m\angle R$. Divide by 2 to show $m\angle A = m\angle R$. This gives you a second pair of angles, and you can prove $\triangle ABC \cong \triangle RST$ by AAS.

## EXERCISE 7.4

1. $\overline{AC} \cong \overline{BC}$ is given. $\overline{CD} \cong \overline{CD}$ by the reflexive property. $\angle ACD \cong \angle BCD$ by the definition of an angle bisector. $\triangle ACD \cong \triangle BCD$ by SAS, and $\angle CAB \cong \angle CBA$ by CPCTC.

2. Because $\overline{PR}$ and $\overline{SQ}$ bisect each other, $PT \cong TR$ and $ST \cong TQ$ by the definition of a bisector. Because vertical angles are congruent, $\angle PTS \cong \angle RTQ$. Then $\triangle PTS \cong \triangle RTQ$ by SAS and $PS \cong QR$ by CPCTC.

3. According to the given information, $\angle BAE \cong \angle DEA$ and $\angle BEA \cong \angle DAE$. By the reflexive postulate, $\overline{AE} \cong \overline{AE}$. $\triangle ABE \cong \triangle EDA$ by ASA and $\overline{AB} \cong \overline{ED}$ by CPCTC.

4. $AB = 10$, $AD = 8$ and $AC = 16$.   Given $BA \cong BC$ and that $\overline{BD}$ bisects $\angle ABC$, show that $\angle ABD \cong \angle CBD$ by the definition of an angle bisector, and $\overline{BD} \cong \overline{BD}$ by the reflexive postulate. $\triangle ABD \cong \triangle CBD$ by SAS and $AD \cong DC$ by CPCTC. Write two equations: $2x + 2y = 10$ and $x + 2y = 4x$. Solve this

system of equations to find $x = 2$ and $y = 3$. Then $AB = 2(2) + 2(3) = 10$, $AD = 2 + 2(3) = 8$ and $AC = 16$.

5. Given that $\overline{RS} \perp \overline{PQ}$ and $\overline{PT} \perp \overline{QR}$, you can conclude that $\angle PSR$ and $\angle RTP$ are right angles because perpendiculars form right angles, and because all right angles are congruent, $\angle PSR \cong \angle RTP$. $\angle SRP \cong \angle TPR$ is also given, and $\overline{PR} \cong \overline{PR}$ by the reflexive postulate. $\triangle SRP \cong \triangle TPR$ by AAS, and $\overline{SP} \cong \overline{TR}$ by CPCTC.

6. $PR = 55$   $SR = 3x+7$ and $PT = 5x - 9$ and in the previous exercise, you proved $SR = PT$, so $3x + 7 = 5x - 9$. Solve to find $x = 8$. $PR = 7x - 1 = 7(8) - 1 = 55$.

7. To prove $\overline{BE} \cong \overline{CE}$, use $\triangle ABE$ and $\triangle DCE$. If you knew that $\triangle ABC \cong \triangle DCB$, which are overlapping triangles, use CPCTC to justify $\angle A \cong \angle D$ and $\overline{AB} \cong \overline{DC}$. Then $\angle AEB \cong \angle DEC$ because vertical angles are congruent, and $\triangle ABE \cong \triangle DCE$ by AAS. Then $\overline{BE} \cong \overline{CE}$ by CPCTC.

## EXERCISE 7.5

1. Three pieces of information are given: $AC \cong BD$, $AE \cong DE$, and $\angle EAC \cong \angle EDB$. This is sufficient to prove $\triangle EAC \cong \triangle EDB$ by SAS, and then $EC \cong EB$ by CPCTC. Notice that $AC = AB + BC$ and $BD = BC + CD$. $AC \cong BD$, so $AC = BD$ by the definition of congruent, and $AB + BCMM = BC + CD$ by substitution. Subtract $BC$ from both sides to show $AB = CD$, so $\overline{AB} \cong \overline{CD}$. This, along with the given information, is sufficient to prove $\triangle AEB \cong \triangle DEC$ by SSS.

2. The given information, $\overline{AG} \cong \overline{FE}$, $\overline{GC} \cong \overline{FC}$, and $\angle AGC \cong \angle EFC$, is sufficient to prove $\triangle AGC \cong \triangle EFC$ by SAS. By CPCTC, $AC \cong DC$. Because vertical angles are congruent, $\angle BCA \cong \angle DCE$, and $\overline{BC} \cong \overline{DC}$ is given. $\triangle ABC \cong \triangle EDC$ by SAS, and $\overline{AB} \cong \overline{ED}$ by CPCTC.

3. It is given that $\overline{TS} \cong \overline{RS}$, $\overline{PS} \cong \overline{QS}$. In addition, $\angle TSP \cong \angle RSQ$ because vertical angles are congruent. $\triangle TSP \cong \triangle RSQ$ by SAS, so $\overline{TP} \cong \overline{RQ}$ and $\angle T \cong \angle R$ by CPCTC. Also given is $\angle RPQ \cong \angle TQP$. $\triangle PTQ \cong \triangle QRP$ by AAS.

4. First prove $\triangle MAB \cong \triangle NCD$. ABCD is a square, which has four congruent sides, so $AB \cong DC$. Because square ABCD is also a rectangle, it has four right angles, so $\angle A$ and $\angle C$ are right angles and all right angles are congruent. Therefore, $\angle A \cong \angle C$. $\overline{AD} \cong \overline{BC}$ as well, with $M$ the midpoint of $\overline{AD}$ and $N$ the midpoint of $\overline{BC}$. $\frac{1}{2} AD = \frac{1}{2} BC$ by the multiplication property of equality, and by the definition of midpoint, $MD = \frac{1}{2} AD$ and $BN = \frac{1}{2} BC$. By substitution, $MD = BN$, or $\overline{MD} \cong \overline{BN}$. $\triangle MAB \cong \triangle NCD$ by SAS. Continue on to prove $\triangle MBD \cong \triangle NDB$. $\overline{MB} \cong \overline{DN}$ by CPCTC. $\angle MBD \cong \angle NDB$ is given, and

$\overline{DB} \cong \overline{DB}$ by the reflexive postulate. $\triangle MBD \cong \triangle NDB$ by SAS.

5. First prove $\triangle ABC \cong \triangle DCB$. These are overlapping triangles so you may want to sketch them separately. The given is $\angle BCA \cong \angle CBD$, and $\angle ABC \cong \angle DCB$, and with $\overline{BC} \cong \overline{BC}$ (reflexive), $\triangle ABC \cong \triangle DCB$ by ASA. By CPCTC, $\overline{AB} \cong \overline{CD}$ and $\overline{AC} \cong \overline{BD}$. $AC = AE + EC$ and $BD = BE + ED$, so $AE + EC = BE + ED$. Because $\angle BCA \cong \angle CBD$, by the Base Angle Converse theorem, $\overline{BE} \cong \overline{EC}$ and so $BE = EC$. By subtracting these equal expressions from each side, $AE + EC = BE + ED$ becomes $AE = ED$ so $\overline{AE} \cong \overline{ED}$. $\triangle AEB \cong \triangle DEC$ by SSS.

6. The given information, $\overline{AF} \cong \overline{DE}$, $\angle FAB \cong \angle EDC$, and $\overline{AB} \cong \overline{CD}$, is sufficient to prove $\triangle AFB \cong \triangle DEC$ by SAS. Then $\overline{BF} \cong \overline{CE}$ by CPCTC. Because perpendiculars form right angles, $\angle BFE$ and $\angle CEF$ are right angles, and because all right angles are congruent, $\angle BFE \cong \angle CEF$. By the reflexive postulate, $\overline{FE} \cong \overline{FE}$, and $\triangle FBE \cong \triangle ECF$ by SAS.

7. The given information that $\angle L \cong \angle R$ and $\overline{LO} \cong \overline{OR}$, together with the vertical angles $\angle LOM \cong \angle ROQ$, provide enough information to conclude that $\triangle LOM \cong \triangle ROQ$ by ASA. Then $\overline{MO} \cong \overline{OQ}$ by CPCTC, and $\angle NOM \cong \angle POQ$ because vertical angles are congruent. $\triangle NOM \cong \triangle POQ$ by AAS and $\overline{MN} \cong \overline{PQ}$ by CPCTC.

---

## CHAPTER 8

# Parallelograms and Other Polygons

## EXERCISE 8.1

1. Because $ABCD$ is a parallelogram, its opposite sides are parallel and congruent, so $\overline{AB} \parallel \overline{CD}$ and $AB = CD$. By the multiplication property of equality, $\frac{1}{2} AB = \frac{1}{2} CD$. Because $M$ and $N$ are midpoints of $\overline{AB}$ and $\overline{CD}$, respectively, $AM = \frac{1}{2} AB$ and $ND = \frac{1}{2} CD$, so $AM = ND$. Because $\overline{AM}$ and $\overline{ND}$ are segments of $\overline{AB}$ and $\overline{CD}$, $\overline{AM} \parallel \overline{ND}$. $AMND$ is a parallelogram because one pair of opposite sides are both parallel and congruent. Therefore, $\overline{MN} \parallel \overline{AD}$ because opposite sides of a parallelogram are parallel.

2. $AE = 3.5$, $AC = 7$, $BE = 1$, and $BD = 2$. Diagonals of a parallelogram bisect each other so $AE = EC$ and $BE = ED$. $AE = EC$ means $x + 2 = 3x - 1$. Simplify and solve to find $x = 1.5$. If $BE = ED$, then $y - 1 = 2y - 3$. Solve to find $y = 2$. $AE = 1.5 + 2 = 3.5$, $AC = 7$, $BE = 2 - 1 = 1$, and $BD = 2$.

3. 68°   Opposite angles of a parallelogram are congruent, so m$\angle Q$ = m$\angle S$, and $11x + 13 = 15x - 23$. Simplify and solve to find $x = 9$. m$\angle Q = 11(9) + 13 = 112°$. m$\angle S = 15(9) - 23 = 112°$. Then m$\angle P$ = m$\angle R = 180° - 112° = 68°$

4. $\angle BAC \cong \angle ECD$ (given) implies $\overline{AB} \parallel \overline{CD}$ because if alternate interior angles congruent, the lines cut by the transversal are parallel. The given information that $\angle CED \cong \angle ECD$ tells us, by the Base Angle Converse theorem, that $\overline{CD} \cong \overline{ED}$. That, along with the given $\overline{ED} \cong \overline{AB}$ and transitivity, means $\overline{CD} \cong \overline{AB}$. Because one pair of sides is parallel and congruent, $ABCD$ is a parallelogram.

5. Given $\angle BAE \cong \angle DCE$, by the Alternate Interior Angle Converse theorem, $\overline{AB} \parallel \overline{CD}$. Then using $\overline{FG}$ as a transversal, $\angle FBA \cong \angle CDG$, because alternate exterior angles are congruent. Given that $\overline{AF} \cong \overline{CG}$ and $\angle AFB \cong CGD$, you

can conclude that $\triangle FAB \cong \triangle GCD$ by AAS. Then $AB \cong CD$ by CPCTC, and $ABCD$ is a parallelogram because one pair of sides are parallel and congruent.

**6.** 32 square centimeters    The perimeter is $P = 2x + 2(x + 3) = 4x + 6 = 26$. Solve to find $x = 5$. The base is $5 + 3 = 8$, and the height is $5 - 1 = 4$. The area is $A = bh = 8(4) = 32$ square centimeters.

**7.** The altitude to the shorter side is twice the altitude to the longer side.    Let $b = $ length of the longer side and $h = $ the length of the altitude to that side. Let $\frac{1}{2}b = $ the length of the shorter side, and $a = $ the length of the altitude to the shorter side. $A = bh = \frac{1}{2}b(a)$. Divide both sides by $b$ to get $h = \frac{1}{2}a$. Multiply by 2 to get $2h = a$. The altitude to the shorter side is twice the altitude to the longer side.

**8.** In any parallelogram, opposite angles are congruent. Therefore, $\angle A \cong \angle C$ in $\square ABCD$, $\angle C \cong \angle E$ in $\square EICJ$, and $\angle E \cong \angle G$ in $\square EFGH$. By the transitive property, $\angle A \cong \angle G$.

---

## EXERCISE 8.2

**1.** Given that $ABCF$ and $DCBE$ are parallelograms, opposite sides are parallel, so $\overline{AF} \parallel \overline{BC}$ and $\overline{BC} \parallel \overline{ED}$. Because $\overline{FE}$ is part of the line containing $\overline{AF}$ and $\overline{ED}$, $\overline{BC} \parallel \overline{FE}$. It is given that $\overline{BC} \cong \overline{FE}$, so $FBCE$ is a parallelogram, because one pair of sides are parallel and congruent. In addition, opposite sides of parallelograms are congruent, so $\overline{AB} \cong \overline{CF}$ and $\overline{BE} \cong \overline{CD}$. Given that $\overline{AB} \cong \overline{DC}$, by transitivity, $\overline{CF} \cong \overline{BE}$. $FBCE$ is a parallelogram with congruent diagonals, so it is a rectangle.

**2.** $HADE$ and $GHCD$ are parallelograms (given) so their opposite sides are parallel. That means $\overline{BD} \parallel \overline{HF}$ and $\overline{HB} \parallel \overline{DF}$. These form the opposite sides of $HBDF$, so $HBDF$ is a parallelogram because both pairs of opposite sides are parallel. To prove $HBDF$ is a rhombus, show that a pair of adjacent sides is congruent. $\angle A \cong \angle C$ is given. Using parallel lines $\overline{AG}$ and $\overline{CE}$ and transversal $\overline{AD}$, $\angle A \cong \angle BDC$ because when parallel lines are cut by a transversal, alternate interior angles are congruent. Using the same reasoning with transversal $\overline{HC}$, $\angle AHB \cong \angle C$. By transitivity, $\angle A \cong \angle BDC \cong \angle AHB \cong \angle C$. By the Base Angle Converse theorem, $\overline{AB} \cong \overline{HB}$ and $\overline{BD} \cong \overline{BC}$. By transitivity $\overline{HB} \cong \overline{BD}$. $HBDF$ is a rhombus.

**3.** Because $ABCD$ is a parallelogram, opposite sides are parallel. $\overline{BC} \parallel \overline{AD}$ and therefore, $\overline{BF} \parallel \overline{ED}$. In addition, opposite sides are congruent, so $\overline{BC} \cong \overline{AD}$ and $BC = AD$. By the multiplicative property of equality, $\frac{1}{2}BC = \frac{1}{2}$

$AD$, and because $F$ and $E$ are the midpoints of $\overline{BC}$ and $\overline{AD}$, respectively, $BF = \frac{1}{2}BC = \frac{1}{2}$ $AD = ED$. $EBFD$ is a parallelogram because one pair of sides are parallel and congruent. By the same logic, $ABFE$ is a parallelogram because $\overline{AE} \cong \overline{BF}$. $\overline{AB} \cong \overline{EF}$ because opposite sides of a parallelogram are congruent, and $\overline{AB} \cong \overline{BD}$ is given, so $\overline{BD} \cong \overline{EF}$ by transitivity and the diagonals of $EBFD$ are congruent, so $EBFD$ is a rectangle. Finally, $\overline{BE} \cong \overline{ED}$ (given), so adjacent sides are congruent, making $EBFD$ a rhombus and therefore a square.

**4.** 1,176 square centimeters    The area of a rhombus $= \frac{1}{2}d_1 d_2 = \frac{1}{2}(42)(56) = 1{,}176$ square centimeters.

**5.** 46.8 cm    Draw the second diagonal of the rhombus and use the Pythagorean theorem in one of the four right triangles to find that half the diagonal is 12, so the diagonals are 10 and 12 and the area of the rhombus is $\frac{1}{2}d_1 d_2 = \frac{1}{2}(12)(10) = 60$. Then $A = bh$ so the height of the rhombus is $\frac{60}{13}$. The area of the rectangle is $A = bh$, so $216 = b$ and $b = 46.8$ cm.

**6.** The rhombus will have a smaller area.    If the square has a perimeter of $4s$, where $s$ is the length of a side, its base and height are both $s$, and its area is $s^2$. If the rhombus also has a perimeter of $4s$, it must have sides of length $s$, but because sides do not meet at right angles, the height is not a side. Draw an altitude from a vertex, and label the section of the base that forms a leg of the right triangle $x$. Use the Pythagorean theorem $x^2 + h^2 = s^2$ to show that $h = \sqrt{s^2 - x^2} < \sqrt{s^2}$. Area of the rhombus would be A = bh = $s\sqrt{s^2 - x^2} < s\sqrt{s^2}$. The rhombus will have a smaller area.

7. Given that $\angle A \cong \angle D$, and with $\angle A \cong \angle BCG$ and $\angle CBG \cong \angle D$ because opposite angles of a parallelogram are congruent, $\angle BCG \cong \angle CBG$, so $\triangle BGC$ is isosceles. If $\triangle BGC$ were equilateral $\overline{BG}$ and $\overline{CG}$ would be congruent to $\overline{BC}$ and therefore congruent to $\overline{BF}$ and $\overline{CE}$, which they clearly are not.

## EXERCISE 8.3

1. 48 square centimeters   The Midsegment theorem says that the midsegment of a triangle is parallel to the third side, and half as long, so $\overline{MN} \parallel \overline{AC}$. Therefore, AMNC has a pair of parallel sides, so it is a trapezoid. The area of $\triangle ABC = \frac{1}{2}(AC)(BD)$=64 square centimeters. The area of $\triangle MBN = \frac{1}{2}(MN)(BE) = \frac{1}{2}(\frac{1}{2}AC)$ $(\frac{1}{2}BD) = \frac{1}{8}(AC)(BD) = \frac{1}{4}(\frac{1}{2}(AC)(BD)) = \frac{1}{4}(64)$ = 16 square centimeters. The area of trapezoid $AMNC$ = the area of $\triangle ABC$ − the area of $\triangle MBN = 64 − 16 = 48$ square centimeters.

2. Given $ABCD$ is a parallelogram, so opposite sides $\overline{BC} \parallel \overline{AD}$ and therefore $\overline{BE} \parallel \overline{AD}$. Because it has one pair of parallel sides, $ABED$ is a trapezoid. $\overline{AB} \cong \overline{CD}$ because opposite sides of a parallelogram are congruent, and $\angle DEC \cong \angle DCE$ is given, and implies that $\overline{CD} \cong \overline{ED}$ by the Base Angle Converse theorem. $ABED$ is an isosceles trapezoid because its non-parallel sides are congruent.

3. Given $\overline{AB} \cong \overline{CD}$, $ABCD$ is an isosceles trapezoid and therefore its base angles are congruent, so $\angle BAD \cong \angle CDA$. $\overline{AD} \cong \overline{AD}$ by the reflexive property and $\triangle ABD \cong \triangle DCA$ by SAS. $\overline{AC} \cong \overline{BD}$ by CPCTC.

4. The trapezoid is not isosceles   $\angle W$ and $\angle X$ are consecutive interior angles along transversal $\overline{WX}$ between $\overline{XY} \parallel \overline{WZ}$. Therefore, $(5x + 2) + (2x + 17) = 180$. Solve to find $7x + 19 = 180$, $7x = 161$, and $x = 23$.

Find the measure of $\angle X = 5(23) + 2 = 117$, and m$\angle Y = 4(23) + 3 = 95$. The base angles are not congruent, so the trapezoid is not isosceles.

5. Rotating quadrilateral $MBCN$ about point $N$ forms parallelogram $AMM'B'$. The base of this parallelogram is $AC + D'B'$, the sum of the top and bottom bases. The height of this parallelogram is half the height of the trapezoid. The area of the trapezoid is half its height times the sum of its bases.

6. a) $A = \frac{1}{2}(10)(14 + 21) = 175$;

   b) $A = \frac{1}{2}(14)(10 + 21) = 217$;

   c) $A = \frac{1}{2}(21)(10 + 14) = 252$

7. 39

$$A = \frac{1}{2}h(b_1 + b_2)$$
$$175 = \frac{1}{2}(7)(11 + x)$$
$$350 = 77 + 7x$$
$$7x = 273$$
$$x = 39$$

8. 10

$$A = \frac{1}{2}h(b_1 + b_2)$$
$$100 = \frac{1}{2}h(8 + 12)$$
$$100 = 10h$$
$$h = 10$$

## EXERCISE 8.4

1. $\overline{AB} \cong \overline{AF} \cong \overline{BC} \cong \overline{CD} \cong \overline{DE} \cong \overline{EF}$
because the hexagon is equilateral. $\angle A \cong \angle C \cong$
$\angle E$ because the hexagon is equiangular.
$\triangle FAB \cong \triangle BCD \cong \triangle DEF$ by SAS, and
$\overline{FB} \cong \overline{BD} \cong \overline{DF}$ by CPCTC. Therefore,
$\triangle BDF$ is equilateral.

2. a) polygon, convex, pentagon; b) not a polygon;
   c) polygon, concave, heptagon; d) polygon,
   convex, quadrilateral; e) polygon, concave,
   heptagon.

3. The figure is not a regular dodecagon, because it
   is not convex and not equiangular.

4. $\overparen{AB} \cong \overparen{CD}$ by construction, so $\angle AOB \cong$
   $\angle COD$ because central angles have the same
   measure as their arcs. $\overline{AO} \cong \overline{BO} \cong \overline{CO} \cong \overline{DO}$
   because all radii of a circle are the same length.
   $\triangle AOB \cong \triangle COD$ by SAS, and $\overline{AB} \cong \overline{CD}$
   by CPCTC.

5. Yes, $\overline{AB} \cong \overline{CD}$ because central angles $\angle AOB$
   and $\angle COD$ are vertical angles and therefore
   congruent. Yes, $\overline{AD} \cong \overline{BC}$ by the same logic
   using $\angle AOD$ and $\angle BOC$. No, $\overline{AB}$ cannot be
   proven congruent to $\overline{AD}$ because $\angle AOB$ is not
   congruent to $\angle AOD$. The inscribed quadrilateral
   is a rectangle but not a square.

6. Draw a radius. Set compass to the length of
   the radius. Mark off six congruent arcs. Connect
   all six marks in sequence to inscribe a hexagon
   and every other mark to form an equilateral
   triangle.

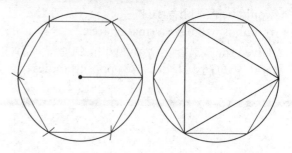

7. Draw a diameter. Construct the perpendicular
   bisector of the diameter. These intersecting
   diameters divide the circle into four congruent
   arcs. Connect the endpoints to form a square.
   Bisect a side of the square and its arc. Repeat for
   each side to divide the circle into eight congruent
   arcs. Connect the endpoints to inscribe a regular
   octagon.

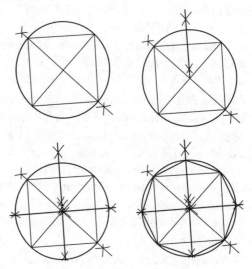

## EXERCISE 8.5

1. 1260°  Sum = $180°(n - 2) = 180°(9 - 2) =$
   1260°

2. 135°  Sum = $180°(n - 2) = 180°(8 - 2) = 1080°$.
   The polygon is regular, so one interior angle is
   $1080° \div 8 = 135°$.

3. 36°  Sum of exterior angles = 360°, and polygon
   is regular with 10 sides, so one exterior angle is
   $360° \div 10 = 36°$.

4. Sum of the interior angles = $180°(n - 2)$. Sum of
   the exterior angles = 360°. If $180°(n - 2) \geq 360°$,

then $n - 2 \geq 2$ and $n \geq 4$. Polygons with 4 or more sides.

**5.** 13   Sum $= 180°(n - 2)$ so solve $1{,}980° = 180°(n - 2)$ or $n - 2 = 11$ and $n = 13$.

**6.** 20 sides   The measure of one angle is $162°$.
Solve $\dfrac{180°(n - 2)}{n} = 162°$ or $180°(n - 2) =$

$162°n$, to get $18n = 360$ of $n = 20$. The polygon has 20 sides.

**7.** $n = 5$   One exterior angle measures $72°$ and the sum of the exterior angles is $360°$, so solve $\dfrac{360°}{n} = 72°$ or $72n = 360$ and $n = 5$.

---

## EXERCISE 8.6

**1.** $54°$   The vertex angle of the triangle is the central angle of the pentagon and measures $360° \div 5 = 72°$. The two base angles are congruent and total $180° - 72° = 108°$. The base angles are congruent, so each measures $108° \div 2 = 54°$.

**2.** 80 cm   The apothem, radius, and half of a side form a right triangle. $a^2 + \left(\tfrac{1}{2}s\right)^2 = r^2$ so $12^2 + \left(\tfrac{1}{2}s\right)^2 = 13^2$. Solve to find that $\tfrac{1}{2}s = 5$ and a side of the octagon $= 10$ cm. The perimeter is 80 cm.

**3.** $5\sqrt{3}$ cm   In a regular hexagon, drawing the radii creates $60°$ central angles and therefore six equilateral triangles, so the radius is also 10 cm. Then $a^2 + \left(\tfrac{1}{2}s\right)^2 = r^2$ becomes $a^2 + 5^2 = 10^2$ becomes $a^2 = 100 - 25 = 75$. The apothem is $5\sqrt{3}$ cm.

**4.** 27.7 square centimeters   In a regular hexagon with perimeter of 48 cm, each side measures 8 cm, the radius is 8 cm and the apothem is $4\sqrt{3}$ cm ($\approx$6.9 cm). The area of one of the six small triangles is $\tfrac{1}{2}bh = \tfrac{1}{2}(8)\left(4\sqrt{3}\right) = 16\sqrt{3} \approx 27.7$ square centimeters.

**5.** 22 square centimeters   A regular pentagon with a perimeter of 40 cm has a side of 8 cm and an apothem of 5.5 cm. The area of one of the small triangles $= \tfrac{1}{2}bh = \tfrac{1}{2}(8)(5.5) = 22$ square centimeters.

**6.** 480 square centimeters   The side is 10 cm so half the side is 5 cm, the perimeter is 80 cm, and the radius is 13 cm, so the apothem is 12 cm. $A = \tfrac{1}{2}aP = \tfrac{1}{2}(12)(80) = 480$ square centimeters.

**7.** 1,725 square centimeters   The side is 15 cm so the perimeter is 150 cm, and the apothem is given as 23 cm. $A = \tfrac{1}{2}aP = \tfrac{1}{2}(23)(150) = 1{,}725$ square centimeters.

**8.** The octagon has the larger area.   A regular hexagon with a side of 8 cm has a perimeter of 48 cm and an apothem of $4\sqrt{3}$ or approximately 6.93 cm. $A = \tfrac{1}{2}aP = \tfrac{1}{2}(6.93)(48) \approx 166.3$ square centimeters. A regular octagon with a side of 6 cm has a perimeter of 48 cm and an apothem of 7.24 cm. $A = \tfrac{1}{2}aP = \tfrac{1}{2}(7.24)(48) \approx 173.8$ square centimeters. The octagon has the larger area.

---

## REVIEW 4

# Congruence, Parallelograms, and Polygons

**1.** At least one pair of congruent sides   There is a great deal of information about congruent angles. Because $\overline{AB} \parallel \overline{DC}$, alternate interior angles are congruent, so $\angle A \cong \angle C$ and $\angle B \cong \angle D$. In addition, $\angle AOB \cong \angle COD$ because vertical angles are congruent. But in order to prove the triangles congruent, it is necessary to have at least one pair of congruent sides.

**2.** Given that $\overline{XZ}$ bisects $\angle WZY$, it is possible to conclude that $\angle WZX \cong \angle YZX$. Given $\overline{XZ}$

bisects ∠WXY, it is possible to conclude that ∠WXZ ≅ ∠YXZ. $\overline{XZ} ≅ \overline{XZ}$ and △WZX ≅ △YZX by ASA.

3. If $\overline{AB} \parallel \overline{DC}$, as given, the ∠A ≅ ∠C by Alternate Interior Angle theorem, and in addition, ∠AOB ≅ ∠COD because vertical angles are congruent. Given that O is the midpoint of $\overline{AC}$, $\overline{AO} ≅ \overline{OC}$, and △AOB ≅ △COD by ASA.

4. $\overline{TR}$ bisects ∠PQR is given and implies ∠PRT ≅ ∠QRT by the definition of bisector. ∠PRQ ≅ ∠QTR is also given and $\overline{TR} ≅ \overline{TR}$. △PTR ≅ △QTR by ASA, and $\overline{PT} ≅ \overline{QT}$ by CPCTC.

5. Given that $\overline{AB} ≅ \overline{CD}$ and ∠ABC ≅ ∠DCB, it is necessary only to add $\overline{BC} ≅ \overline{BC}$ (reflexive) and △ABC ≅ △DCB by SAS. By CPCTC, $\overline{AC} ≅ \overline{DB}$ and ∠A ≅ ∠D. Because vertical angles are congruent, ∠BEA ≅ ∠CED, and then △BEA ≅ △CED by AAS. Finally, $\overline{AE} ≅ \overline{ED}$ by CPCTC.

6.

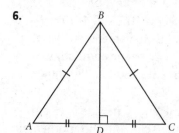

△ABC is isosceles with $\overline{AB} ≅ \overline{BC}$ and $\overline{BD}$ is the perpendicular bisector of $\overline{AC}$. $\overline{BD} ≅ \overline{BD}$. ∠ADB and ∠CDB are right angles formed by the perpendiculars, and because all right angles are congruent, ∠ADB ≅ ∠CDB. $\overline{AD} ≅ \overline{DC}$ by the definition of bisector. △ABD ≅ △CDB by SAS, and ∠ABD ≅ ∠CDB by CPCTC. By the definition of angle bisector, $\overline{BD}$ bisects ∠ABC.

7. Rectangle and square

8. Rhombus and square

9. Parallelogram, rectangle, rhombus, square

10. Rhombus and square

11. Trapezoid, isosceles trapezoid, parallelogram, rectangle, rhombus, square

12. If M is the midpoint of $\overline{AC}$ and the midpoint of $\overline{BD}$, then diagonals $\overline{AC}$ and $\overline{BD}$ bisect each other, and ABCD is a parallelogram. In a parallelogram, opposite sides are congruent, so $\overline{AD} ≅ \overline{BC}$. Alternately, $\overline{AM} ≅ \overline{MC}$ and $\overline{DM} ≅ \overline{MB}$ by the definition of midpoint

and ∠AMD ≅ ∠BMC because vertical angles are congruent. △AMD ≅ △BMC by SAS and $\overline{AD} ≅ \overline{BC}$ by CPCTC.

13. 10   In a parallelogram, opposite angles are congruent and consecutive angles are supplementary. Use those facts to write two equations, each involving x and y. Solve the system to find the values of x and y.

$$10y + x = 5y + 7x$$
$$\underline{(2x + 2y + 6) + (10y + x) = 180}$$
$$-6x + 5y = 0$$
$$\underline{3x + 12y = 174}$$
$$-6x + 5y = 0$$
$$\underline{6x + 24y = 348}$$
$$29y = 348$$
$$y = 12$$
$$-6x + 5(12) = 0$$
$$-6x = -60$$
$$x = 10$$

14. In a parallelogram, opposite angles are congruent. If ABCD is a parallelogram, ∠A ≅ ∠C. If AEFG is a parallelogram, ∠A ≅ ∠F. By transitivity, ∠C ≅ ∠F. ∠C and ∠D are consecutive angles of a parallelogram, and so they are supplementary, which means m∠C + m∠D = 180°. ∠C ≅ ∠F, so m∠C = m∠F and by substituting, m∠F + m∠D = 180°. ∠F and ∠D are supplementary.

15.

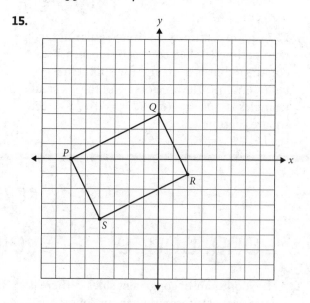

First, verify that *PQRS* is a parallelogram by checking that opposite sides are parallel. Find the slopes and show that opposite sides have the same slope. For $\overline{PQ}$,

$$m = \frac{y_2 - y_1}{x_2 - x_1} = \frac{3-0}{0+6} = \frac{3}{6} = \frac{1}{2}.$$

For $\overline{RS}$, $m = \frac{y_2 - y_1}{x_2 - x_1} = \frac{-4+1}{-4-2} = \frac{-3}{-6} = \frac{1}{2}.$

For $\overline{QR}$, $m = \frac{y_2 - y_1}{x_2 - x_1} = \frac{-1-3}{2-0} = \frac{-4}{2} = -2.$

For $\overline{PS}$, $m = \frac{y_2 - y_1}{x_2 - x_1} = \frac{-4-0}{-4+6} = \frac{-4}{2} = -2.$

Opposite sides have the same slope, and so are parallel, therefore *PQRS* is a parallelogram. To show that *PQRS* is a rectangle, show that adjacent sides are perpendicular. It is only necessary to show that there is one right angle.

For $\overline{PQ}$, $m = \frac{y_2 - y_1}{x_2 - x_1} = \frac{3-0}{0+6} = \frac{3}{6} = \frac{1}{2}.$

For $\overline{QR}$, $m = \frac{y_2 - y_1}{x_2 - x_1} = \frac{-1-3}{2-0} = \frac{-4}{2} = -2.$

Because the slopes of $\overline{PQ}$, and $\overline{QR}$ multiply to $\frac{1}{2}(-2) = -1$, $\overline{PQ} \perp \overline{QR}$, so $\angle PQR$ is a right angle. Because it is a parallelogram with a right angle, *PQRS* is a rectangle.

**16.** 10 sides   Solve $\frac{180°(n-2)}{n} = 144$ to find

$144n = 180(n-2)$ and $n = 10$. The polygon has 10 sides.

**17.** 128.6°   In a regular heptagon, each interior angle measures $\frac{5(180)}{7} = 128\frac{4}{7}° \approx 128.6°.$

**18.** 2,340°   In a polygon with 15 sides, the total of the interior angles is $180°(n-2) = 13(180) = 2,340°.$

**19.**

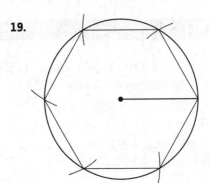

**20.** 374.1 square centimeters   If the radius of the circle is 12 cm, the radius of the hexagon is 12 cm, the side of the hexagon is 12 cm, divided into 2 segments of 6 cm by the apothem.

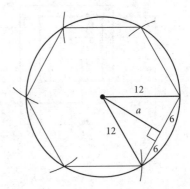

Because the right triangle is a 30°-60°-90° right triangle (or by using the Pythagorean theorem) the apothem is $6\sqrt{3}$ cm.
$A = \frac{1}{2}aP = \frac{1}{2}(6\sqrt{3})(72) = 216\sqrt{3} \approx 374.1$ square centimeters.

CHAPTER 9

# Similarity

## EXERCISE 9.1

**1.** 32.2 cm 6 cm × 0.7 = 4.2 cm. 17 cm × 0.7 = 11.9 cm. P = 2(4.2) + 2(11.9) = 8.4 + 23.8 = 32.2 cm.

**2.** R(3, −4)→R′(6, −8). S(5, −1)→S′(10, −20. T(2, 4)→T′(4, 8).

**3.** A(−2, 2)→A′(−3, 3). B(4, 2)→B′(6, 3). C(4, −4)→C′(6, −6). D(−2, −4)→D′(−3, −6).

**4.** $k = \frac{2}{3}$ To carry A′B′C′D′ back onto ABCD use a scale factor of k, where k(−3) = −2, so $k = \frac{2}{3}$.

**5.**

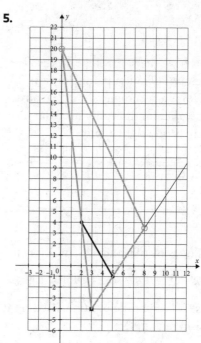

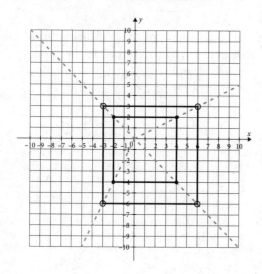

| R(3, −4) | Translate to move R to the origin. | (0, 0) | Dilate by a factor of 2.5 | (0, 0) | Translate R to (3, −4) | (3, −4) |
|---|---|---|---|---|---|---|
| S(5, −1) | | (2, 3) | | (5, 7.5) | | (8, 3.5) |
| T(2, 4) | | (−1, 8) | | (−2.5, 20) | | (0.5, 16) |

**6.**

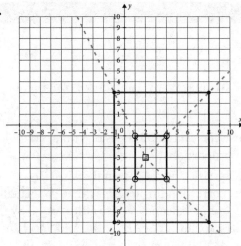

| A(−1, 3) | Translate to move P(2, −3) to the origin. | (−3, 6) | Dilate by a factor of $\frac{1}{3}$ | (−1, 2) | Translate P to (2, −3) | (1, −1) |
|---|---|---|---|---|---|---|
| B(8, 3) | | (6, 6) | | (2, 2) | | (4, −1) |
| C(8, −9) | | (6, −6) | | (2, −2) | | (4, −5) |
| D(−1, −9) | | (−3, −6) | | (−1, −2) | | (1, −5) |

**7.** A negative scale factor would represent a dilation combined with a rotation of 180° about the origin.

---

## EXERCISE 9.2

**1.** If $\triangle XYZ \sim \triangle RTS$, then $\angle X \cong \angle R$, $\angle Y \cong \angle T$ and $\angle Z \cong \angle S$. $\overline{XY}$ corresponds to $\overline{RT}$, $\overline{YZ}$ to $\overline{TS}$, and $\overline{XZ}$ to $\overline{TS}$.

**2.** The two statements involve the same triangles, but establish different correspondences. In the first, X corresponds to R, Y to T and Z to S. In the second X again corresponds to R, but Y corresponds to S, and Z to T.

**3.** If m$\angle A = 43°$ and m$\angle B = 112°$, then m$\angle C = 180° − (43° + 112°) = 25°$. If m$\angle X = 25°$ and m$\angle Y = 43°$, then m$\angle Z = 180° − (25° + 43°) = 112°$. $\triangle ABC \sim \triangle YZX$

**4.** If a correspondence exists, it will map shortest side to shortest side (12→6), middle to middle (22→12), and longest to longest (30→18). But these do not show the same scale factor so the triangles are not similar.

**5.** Check scale factors: $\frac{18}{27} = \frac{2}{3}$, $\frac{25}{37.5} = \frac{50}{75} = \frac{2}{3}$, and $\frac{40}{60} = \frac{2}{3}$. $\triangle ARM \sim \triangle GEL$ with a scale factor of $\frac{2}{3}$.

**6.** Locate the largest and smallest angle of each polygon and the longest and shortest side of each. Rotate and/or reflect the polygons until these coincide, largest with largest, smallest with smallest. *CLAMP ~ NDTRE*

**7.** Perimeter: 2.4, Area: 5.76 The original is 6 by 10, so it has $P = 2(6) + 2(10) = 32$, and $A = 6(10) = 60$. The dilated rectangle is 14.4 by 24, so it has $P = 2(14.4) + 2(24) = 76.8$, and $A = 14.4(24) = 345.6$. The perimeter was enlarged by a factor of $\frac{76.8}{32} = 2.4$ and the area was enlarged by $\frac{345.6}{60} = 5.76 = (2.4)^2$.

## EXERCISE 9.3

**1.** By the Triangle Sum theorem, $m\angle A + m\angle B + m\angle C = 180°$ and $m\angle X + m\angle Y + m\angle Z = 180°$, so $m\angle A + m\angle B + m\angle C = m\angle X + m\angle Y + m\angle Z$. Given that $m\angle A = m\angle X$ and $m\angle B = m\angle Y$, subtract $m\angle A + m\angle B$ from one side and $m\angle X + m\angle Y$ from the other, leaving $m\angle C = m\angle Z$.

**2.** $\triangle ABC \sim \triangle DEC$ by AA Given $\overline{AB} \parallel \overline{DE}$, cut by a transversal, alternate interior angles $\angle A \cong \angle D$ and alternate interior angles $\angle B \cong \angle E$. $\triangle ABC \sim \triangle DEC$.

**3.** $\triangle ABC \sim \triangle DEC$ by AA Given that $\overline{AB} \parallel \overline{DF}$, when these parallel lines are cut by a transversal, corresponding angles are congruent, so $\angle A \cong \angle EDC$. $\angle BCA$ and $\angle ECD$ are in fact the same angle, so $\angle BCA \cong \angle ECD$. $\triangle ABC \sim \triangle DEC$ by AA.

**4.** $\triangle CDE \sim \triangle BDF$ by AA Given $\overline{CE} \parallel \overline{BF} \parallel \overline{AG}$, cut by a transversal, corresponding angles are congruent, so $\angle DCE \cong \angle DBF$ and $\angle DEC \cong \angle DFB$. $\triangle CDE \sim \triangle BDF$ by AA.

**5.** $\dfrac{PQ}{SR} = \dfrac{PT}{TS}$ Given that $\overline{RT} \perp \overline{PS}$, $\angle QTP$ and $\angle RTS$ are right angles because perpendiculars form right angles, and because all right angles are congruent, $\angle QTP \cong \angle RTS$. That $\angle PQT \cong \angle SRT$ is given and so $\triangle PQT \sim \triangle SRT$. Then $\dfrac{PQ}{SR} = \dfrac{PT}{TS}$ because corresponding sides of similar triangles are in proportion.

**6.** $\dfrac{WY}{ZY} = \dfrac{WX}{ZV}$ Given $\angle X \cong \angle Y$, and $\angle XYW \cong \angle VYZ$ because vertical angles are congruent, $\triangle XYW \sim \triangle VYZ$ by AA. Then $\dfrac{WY}{ZY} = \dfrac{WX}{ZV}$ because corresponding sides of similar triangles are in proportion.

**7.** $\triangle ABC \sim \triangle JKL$ by AA Given that $\triangle ACD \sim \triangle JLM$, corresponding angles will be congruent, so $\angle C \cong \angle L$. $AB$ and $JK$ are altitudes, so by the definition of altitude, $AB \perp DC$ and $JK \perp ML$. Perpendiculars form right angles $\angle ABC$ and $\angle JKL$, and all right angles are congruent so $\angle ABC \cong \angle JKL$. $\triangle ABC \sim \triangle JKL$ by AA.

**8.** $RS = \frac{1}{2}PQ$ $\overline{PQ} \cong \overline{QR}$ and $\overline{RS} \cong \overline{ST}$ are given. By the Base Angle theorem, $\angle QPR \cong \angle QRP$ and $\angle SRT \cong \angle STR$. If $\overline{PQ} \parallel \overline{ST}$ (given), then $\angle QPR \cong \angle SRT$ because alternate interior angles are congruent, and by transitivity, $\angle QRP \cong \angle SRT \cong \angle STR \cong \angle QPR$. $\triangle PQR \sim \triangle RST$ by AA. Then $\dfrac{TR}{PR} = \dfrac{RS}{PQ}$ because corresponding sides of similar triangles are in proportion. Because $T$ is the midpoint of $\overline{PR}$, $TR = \frac{1}{2}PR$, or $\dfrac{TR}{PR} = \dfrac{1}{2}$, and by transitivity $\dfrac{RS}{PQ} = \dfrac{1}{2}$ or $RS = \frac{1}{2}PQ$.

## EXERCISE 9.4

**1.** $x = 4$ Cross-multiply $\dfrac{x}{7} = \dfrac{12}{21}$ to get $21x = 84$ and $x = 4$.

**2.** $x = 5\frac{2}{11} \approx 5.\overline{18}$ Cross-multiply $\dfrac{11}{18} = \dfrac{5}{x+3}$ to get $11x + 33 = 90$, and $11x = 57$, so $x = 5\frac{2}{11} \approx 5.\overline{18}$.

**3.** The proportion $\dfrac{a}{b} = \dfrac{c}{d}$ cross-multiplies to $ad = bc$. Variant $\dfrac{b}{a} = \dfrac{d}{c}$ becomes $ad = bc$ directly, $\dfrac{a+b}{b} = \dfrac{c+d}{d}$ becomes $ad + bd = bc + bd$. Subtract $bd$ from both sides and $ad = bc$.

Similarly, $\dfrac{a-b}{b} = \dfrac{c-d}{d}$ becomes $ad - bd = bc - bd$. Add $bd$ to both sides and $ad = bc$.

**4.** $x = 16\frac{2}{7} \approx 16.29$ $\dfrac{AB}{XY} = \dfrac{BC}{YZ}$ becomes $\dfrac{14}{49} = \dfrac{x}{57}$. Cross-multiply to get $49x = 798$ and $x = 16\frac{2}{7} \approx 16.29$.

**5.** $4$ $\dfrac{AB}{A'B'} = \dfrac{9}{4}$ so $A'B' = 4$.

**6.** $X'Z' = 71.5$ $\dfrac{X'Y'}{XY} = \dfrac{13}{20}$ becomes $\dfrac{156}{XY} = \dfrac{13}{20}$. Cross-multiply and solve to get

$XY = \dfrac{20(156)}{13} = 240$. $\dfrac{X'Z'}{XZ} = \dfrac{13}{20}$ becomes

$\dfrac{X'Z'}{110} = \dfrac{13}{20}$. Cross-multiply and solve to get

$X'Z' = \dfrac{110(13)}{20} = 71.5$.

**7.** 5:12 Cross-multiply $\dfrac{x}{12} = \dfrac{5}{x+7}$

to get $x^2 + 7x = 60$. Solve by

factoring. $x^2 + 7x - 60 = 0$ factors to $(x + 12)(x - 5) = 0$, and when each factor is set equal to zero, the solutions are x = −12, which is rejected as unreasonable, and x = 5. The proportion becomes $\dfrac{5}{12} = \dfrac{5}{5+7}$. The scale factor is 5:12.

## EXERCISE 9.5

**1.** If $\dfrac{CX}{CA} = \dfrac{CY}{CB}$, then $\dfrac{CX}{CA - CX} = \dfrac{CY}{CB - CY}$.

$CA = CX + XA$, so $CA - CX = XA$. Likewise,

$CB = CY + YB$, so $CB - CY = YB$. The

proportion becomes $\dfrac{CX}{XA} = \dfrac{CY}{YB}$.

**2.** In $\triangle CWV$, $\overline{XY} \parallel \overline{WV}$ and $\dfrac{CX}{XW} = \dfrac{CY}{YV}$ or

$\boxed{\dfrac{XW}{CX} = \dfrac{YV}{CV}}$. In $\triangle CAB$, $\overline{XY} \parallel \overline{AB}$ and

$\dfrac{CX}{XA} = \dfrac{CV}{YB}$ or $\dfrac{XA}{CX} = \dfrac{YB}{CY}$. Rewrite $XA$

and $YB$ to get $\dfrac{XW + WA}{CX} = \dfrac{YV + VB}{CY}$

or $\dfrac{XW}{CX} + \dfrac{WA}{CX} = \dfrac{YV}{CY} + \dfrac{VB}{CY}$. Because

$\dfrac{XW}{CX} = \dfrac{YV}{CV}$ from above, subtract:

$\dfrac{\cancel{XW}}{\cancel{CX}} + \dfrac{WA}{CX} = \dfrac{\cancel{YV}}{\cancel{CY}} + \dfrac{VB}{CY}$ and $\dfrac{WA}{CX} = \dfrac{VB}{CY}$ or

$\boxed{\dfrac{CX}{WA} = \dfrac{CY}{VB}}$. Finally, in $\triangle CAB$, $\overline{WV} \parallel \overline{AB}$ and

$\dfrac{CW}{WA} = \dfrac{CV}{VB}$. Rewrite: $\dfrac{CX + XW}{WA} = \dfrac{CY + YV}{VB}$

or $\dfrac{CX}{WA} + \dfrac{XW}{WA} = \dfrac{CY}{VB} + \dfrac{YV}{VB}$. Subtract:

$\dfrac{\cancel{CX}}{\cancel{WA}} + \dfrac{XW}{WA} = \dfrac{\cancel{CY}}{\cancel{VB}} + \dfrac{YV}{VB}$ or $\dfrac{XW}{WA} = \dfrac{YV}{VB}$.

**3.** $\dfrac{XY}{AB}$ is $\dfrac{\text{whole side of } \triangle CXY}{\text{whole side of } \triangle CAB}$ so $\dfrac{CX}{CA}$ is correct.

**4.** $x = \dfrac{20}{7}$ Let $x = WZ$ and $5 - x = ZY$. Because

$\overline{XZ}$ bisects $\angle X$, $\dfrac{WZ}{WX} = \dfrac{ZY}{XY}$. Substitute

to get $\dfrac{x}{4} = \dfrac{5 - x}{3}$ and cross-multiply to

get $3x = 20 - 4x$. Solve to get $7x = 20$ and

$x = \dfrac{20}{7} = 2\dfrac{6}{7} \approx 2.86$.

**5.** $BC = 25$ cm and $AB = 40$ cm   Let $x = BC$

and $x + 15 = AB$. Then $\dfrac{32}{x + 15} = \dfrac{20}{x}$. Cross-

multiply to get $32x = 20x + 300$. Solve to get

$12x = 300$ and $x = 25$ cm. $BC = 25$ cm and $AB = 40$ cm.

**6.** Focus on $\triangle ABX$. In this triangle, $\overline{BP}$ bisects $\angle B$ and divides $\overline{AX}$ so that the segments $\overline{AP}$ and $\overline{PX}$ are proportional to the sides of the triangle $\overline{AB}$ and $\overline{BX}$. Therefore, $\dfrac{AP}{AB} = \dfrac{PX}{BX}$.

## EXERCISE 9.6

**1.** $x = 4$  $\dfrac{AD}{BD} = \dfrac{BD}{DC}$ becomes $\dfrac{x}{14} = \dfrac{14}{49}$. Cross-multiply and solve. $49x = 196$ and $x = 4$.

**2.** $x = 9$  $\dfrac{AD}{AB} = \dfrac{AB}{AC}$ becomes $\dfrac{3}{6} = \dfrac{6}{3 + x}$. Cross-multiply and solve. $9 + 3x = 36$ and $x = 9$.

**3.** $DC = 18$ and $AC = 26$  $\dfrac{AB}{BD} = \dfrac{BD}{DC}$ becomes $\dfrac{8}{12} = \dfrac{12}{x}$. Cross-multiply and solve. $8x = 144$ and $x = 18$. $DC = 18$ and $AC = AD + DC = 8 + 18 = 26$.

**4.** $x = 9$  $\dfrac{DC}{BC} = \dfrac{BC}{AC}$ becomes $\dfrac{x}{15} = \dfrac{15}{25}$. Cross-multiply and solve. $25x = 225$ and $x = 9$.

**5.** $BC = \sqrt{1,280} \approx 35.8$  $\dfrac{8}{16} = \dfrac{16}{DC}$ becomes $8DC = 256$ and $DC = 32$. $AC = AD + DC = 8 +$

$32 = 40$. $\dfrac{AD}{AB} = \dfrac{AB}{AC}$ becomes $\dfrac{8}{AB} = \dfrac{AB}{40}$. Cross-multiply and solve. $(AB)^2 = 320$ and $AB = \sqrt{320} \approx 17.9$. Finally, $\dfrac{DC}{BC} = \dfrac{BC}{AC}$ becomes $\dfrac{32}{BC} = \dfrac{BC}{40}$ so $(BC)^2 = 1,280$ and $BC = \sqrt{1,280} \approx 35.8$.

**6.** $\dfrac{x}{a} = \dfrac{a}{c}$ becomes $a^2 = xc$ and $\dfrac{y}{b} = \dfrac{b}{c}$ becomes $b^2 = yc$.

$a^2 + b^2 = xc + yc = c(x + y) = c(c) = c^2$

**7.** The Pythagorean theorem.

---

# CHAPTER 10

# Right Triangle Trigonometry

---

## EXERCISE 10.1

**1.** $\triangle PQR \sim \triangle ZXY$

**2.** $\dfrac{DO}{CA} = \dfrac{OG}{AT} = \dfrac{DG}{CT}$

**3.** 25.5 cm.  $\dfrac{AR}{LE} = \dfrac{RM}{EG}$ becomes $\dfrac{22}{33} = \dfrac{17}{x}$. Cross-multiply and solve to get $x = 25.5$ cm.

**4.** 5:1  $\dfrac{PQ}{ML} = \dfrac{QR}{LK}$ becomes $\dfrac{21}{y} = \dfrac{25}{5}$. Cross-multiply and solve to get $y = 4.2$ in. The scale factor is 5:1.

**5.** 6.5

$c^2 = a^2 + b^2 = 21^2 + 25^2 = 441 + 625 = 1,066$

$c = \sqrt{1,066} \approx 32.6$

$c^2 = a^2 + b^2 = (4.2)^2 + 5^2 = 17.64 + 25 = 42.64$

$c = \sqrt{42.64} \approx 6.5$

**6.** $\sqrt{3}$  $\dfrac{\text{opposite}}{\text{adjacent}} = \dfrac{\sqrt{3}}{1} = \dfrac{3\sqrt{3}}{3} = \dfrac{5\sqrt{3}}{5} = \sqrt{3}$

**7.** $\dfrac{\sqrt{2}}{2}$  $\dfrac{\text{opposite}}{\text{hypotenuse}} = \dfrac{1}{\sqrt{2}} = \dfrac{4}{4\sqrt{2}} = \dfrac{8}{8\sqrt{2}} = \dfrac{\sqrt{2}}{2}$

## EXERCISE 10.2

**1.** $\sin \angle R = \dfrac{5}{13}$

**2.** $\cos \angle T = \dfrac{5}{13}$

**3.** $\sec \angle T = \dfrac{13}{5} = 2.6$

**4.** $\cot \angle R = \dfrac{12}{5} = 2.4$

**5.** $\cot \angle T = \dfrac{5}{12}$

**6.** $\tan \angle R = \dfrac{5}{12}$

**7.** By definition, $\sin \angle X = \dfrac{\text{side opposite } \angle X}{\text{hypotenuse}}$ but if $\angle X$ is the right angle, the hypotenuse is the also

the opposite side. $\cos \angle X = \dfrac{\text{side adjacent to } \angle X}{\text{hypotenuse}}$ but if $\angle X$ is the right angle both legs are adjacent to $\angle X$. $\tan \angle X = \dfrac{\text{opposite}}{\text{adjacent}}$ has the same ambiguity. The current definitions are not workable for right angles.

**8.** a) $\sin 30° = 0.5$
   b) $\cos 45° \approx 0.707$
   c) $\tan 82° \approx 7.115$
   d) $\cos 11° \approx 0.982$
   e) $\sin 75° \approx 0.966$

## EXERCISE 10.3

**1.** a) $\sin 78° \approx 0.978$
   b) $\cos 15° \approx 0.966$
   c) $\cos 43° \approx 0.731$
   d) $\tan 62° \approx 1.881$
   e) $\sin 21° \approx 0.358$

**2.** 19 cm   $\cos(30°) = \dfrac{XY}{38}$, so $XY = 38 \cos(30°) \approx$ 32.9 cm. Then $\sin(30°) = \dfrac{YZ}{38}$, so $YZ = 38 \sin(30°) = 19$ cm.

**3.** 35.7 cm   $\sin(79°) = \dfrac{35}{AC}$, so $AC \sin(79°) = 35$ and $AC = \dfrac{35}{\sin(79°)} \approx 35.7$ cm.

**4.** 16.97 inches   In an isosceles right triangle with a leg of 12 inches, both legs measure 12 inches,

and the length of the hypotenuse could be found, using the Pythagorean theorem, as $\sqrt{288} = 12\sqrt{2} \approx 16.97$ inches. Working instead with trig ratios, $\sin(45°) = \dfrac{12}{h}$ can be solved to find $h = \dfrac{12}{\sin(45°)} \approx \dfrac{12}{0.7071} \approx 16.97$ inches.

**5.** 13.4 inches   $\tan(40°) = \dfrac{YZ}{16}$ so $YZ = 16 \tan(40°) \approx 13.4$ inches.

**6.** 11.6 inches   $\cos(15°) = \dfrac{AD}{12}$ so $AD = 12 \cos(15°) \approx 11.6$ inches.

**7.** 7.3 meters   $\cos(47°) = \dfrac{5}{XZ}$ so $XZ \cos(47°) = 5$ and $XZ = \dfrac{5}{\cos(47°)} \approx 7.3$ meters.

## EXERCISE 10.4

**1.** a) $\tan^{-1}(2.904) \approx 71°$

b) $\cos^{-1}(0.777) \approx 39°$

c) $\sin^{-1}(0.819) \approx 55°$

d) $\cos^{-1}(0.139) \approx 82°$

e) $\tan^{-1}(1.000) = 45°$

**2.** 57.8°   $\tan \angle A = \dfrac{62}{39}$ so $m\angle A = \tan^{-1}\left(\dfrac{62}{39}\right) = 57.8°$

**3.** 37.5°   $\sin \angle L = \dfrac{14}{23}$ so $m\angle L = \sin^{-1}\left(\dfrac{14}{23}\right) = 37.5°$

**4.** 41.2°   $\tan \angle Z = \dfrac{35}{40}$ so $m\angle Z = \tan^{-1}\left(\dfrac{35}{40}\right) \approx 41.2°$

**5.** 30°   $\sin \angle C = \dfrac{10}{20}$ so $m\angle C = \sin^{-1}\left(\dfrac{10}{20}\right) = 30°$

**6.** 45°   $\sin^{-1}\left(\dfrac{17}{17\sqrt{2}}\right) = \sin^{-1}\left(\dfrac{\sqrt{2}}{2}\right) = 45°$

**7.** 53.1°   $\tan \angle A = \dfrac{3}{4}$ so $m\angle A = \tan^{-1}\left(\dfrac{3}{4}\right) \approx 36.9°$

$\tan \angle C = \dfrac{4}{3}$ so $m\angle C = \tan^{-1}\left(\dfrac{4}{3}\right) \approx 53.1°$

## EXERCISE 10.5

**1.** $\csc \angle A = \dfrac{8}{3}$

**2.** $\cos \angle R = \dfrac{7}{12}$

**3.** $\cot \angle X = \dfrac{1}{3.1} = \dfrac{10}{31} \approx 0.3226$

**4.** $\sin \angle L = \dfrac{\text{opposite}}{\text{hypotenuse}}$ and $\cos \angle L = \dfrac{\text{adjacent}}{\text{hypotenuse}}$.

$\dfrac{\sin \angle L}{\cos \angle L} = \dfrac{\text{opposite/hypotenuse}}{\text{adjacent/hypotenuse}}$ Simplify:

$\dfrac{\sin \angle L}{\cos \angle L} = \dfrac{\text{opposite}}{\text{hypotenuse}} \div \dfrac{\text{adjacent}}{\text{hypotenuse}}$

$= \dfrac{\text{opposite}}{\text{hypotenuse}} \cdot \dfrac{\text{hypotenuse}}{\text{adjacent}}$

$= \dfrac{\text{opposite}}{\text{adjacent}} = \tan \angle L$

**5.** $\sqrt{3}$   $\tan(60°) = \dfrac{\sin(60°)}{\cos(60°)} = \dfrac{\sqrt{3}/2}{1/2} = \sqrt{3}$

**6.** $\dfrac{\sqrt{2}}{2}$   $\tan(45°) = \dfrac{\sin(45°)}{\cos(45°)}$ so $1 = \dfrac{\sqrt{2}/2}{\cos(45°)}$

implies $\cos(45°) = \dfrac{\sqrt{2}}{2}$

**7.** 0.766   $\dfrac{\cos(40°)}{\sin(40°)} = \cot(40°)$ so $\dfrac{\cos(40°)}{0.643} = 1.192$

and $\cos(40°) = (1.192)(0.643) \approx 0.766$

**8.** $\cot(15°) = 3.732$

**9.** $\sin(30°) = 0.5$

**10.** 0.267   $\cot(75°) = \dfrac{1}{\tan(75°)} = \dfrac{1}{3.732} \approx 0.267$

**11.** a) 0.954

b) 0.314

c) 3.18

d) 1.048

e) 3.333...

f) 0.954

g) 0.3

h) 3.18

i) 0.314

j) 1.048

a) $\cos(\angle A)$: $\sin^2 \angle A + \cos^2 \angle A = 1$

$(0.3)^2 + \cos^2 \angle A = 1$

$\cos^2 \angle A = 1 - 0.09 = 0.91$

$\cos \angle A = \sqrt{0.91} \approx 0.954$

b) $\tan(\angle A) = \dfrac{0.3}{0.954} \approx 0.314$

c) $\cot(\angle A) = \dfrac{0.954}{0.3} \approx 3.180$

d) $\sec(\angle A) = \dfrac{1}{0.954} \approx 1.048$

e) $\csc(\angle A) = \dfrac{1}{0.3} = \dfrac{10}{3} = 3.\overline{3}$

f) $\sin(\angle C) = \cos(\angle A) = 0.954$
g) $\cos(\angle C) = \sin(\angle A) = 0.3$
h) $\tan(\angle C) = \cot(\angle A) = 3.18$
i) $\cot(\angle C) = \tan(\angle A) = 0.314$
j) $\sec(\angle C) = \csc(\angle A) = 3.\overline{3}$
k) $\csc(\angle C) = \sec(\angle A) = 1.048$

## EXERCISE 10.6

**1.** 0.951   $\sin^2(18°) + \cos^2(18°) = 1$, so

$\cos^2(18°) = 1 - \sin^2(18°) = 1 - (0.309)^2$.

$\cos(18°) = \sqrt{0.9045} \approx 0.951$

**2.** 0.454   $\sin^2(27°) + \cos^2(27°) = 1$, so $\sin^2(27°) =$

$1 - \cos^2(27°) = 1 - (0.891)^2 = 0.206119$.

$\sin(27°) = \sqrt{0.206119} \approx 0.454$

**3.** 1.221   $\sec^2(35°) = \tan^2(35°) + 1 = (0.700)2 +$

$1 = 1.49$ so $\sec(35°) = \sqrt{1.49} \approx 1.221$

**4.** 1.11   $\cot^2(42°) + 1 = \csc^2(42°)$ so $\cot^2(42°) =$

$sc^2(42°) - 1 = (1.494)^2 - 1 \approx 1.232$ and

$\cot(42°) = \sqrt{1.232} \approx 1.11$

**5.** 0.515   To find $\cos(59°)$, first find $\sec(59°)$.
$\sec^2(59°) = \tan^2(59°) + 1 = (1.664)^2 + 1 \approx 3.769$

and $\sec(59°) = \sqrt{3.769} \approx 1.9414$ . Then

$\cos(59°) = \dfrac{1}{\sec(59°)} = \dfrac{1}{1.9414} \approx 0.515$

**6.** 1.803   If $\cos(61°) = 0.485$, then $\sec(61°) =$

$\dfrac{1}{0.485} \approx 2.0619$. Then $\tan^2(61°) = \sec^2(61°) - 1 =$

$(2.0619)^2 - 1 \approx 3.2514$ so $\tan(61°) =$

$\sqrt{3.2514} \approx 1.803$

**7.** 0.966   Given $\tan(75°)$, to find $\sin(75°)$, first find
$\cot(75°)$, then find $\csc(75°)$, and finally $\sin(75°)$. If

$\tan(75°) = 3.732$, then $\cot(75°) = \dfrac{1}{3.732} \approx 0.268$.

$\csc^2(75) = 1 + \cot^2(75°) \approx 1.0718$ and

$\csc(75°) = \sqrt{1.0718} \approx 1.035$. Finally,

$\sin(75°) = \dfrac{1}{\csc(75°)} = \dfrac{1}{1.035} \approx 0.966$.

## EXERCISE 10.7

**1.** 525.3 feet   $\tan(35°) = \dfrac{x}{250}$ so $x = 250$

$\tan(35°) \approx 175.1$ yards or 525.3 feet.

**2.** 27.8 feet   $\tan(48°) = \dfrac{x}{25}$ so $x = 25 \tan(48°) \approx$

27.8 feet.

**3.** 37.9°   $\tan(x°) = \dfrac{7}{9}$ so $x° = \tan^{-1}\left(\dfrac{7}{9}\right) \approx 37.9°$.

**4.** 1,343.8 feet   $\tan(7°) = \dfrac{165}{x}$ so $x = \dfrac{165}{\tan(7°)} \approx$

1,343.8 feet or about a quarter of a mile.

**5.** 5.2 feet    $\sin(12°) = \dfrac{x}{25}$ so $x = 25\sin(12°) \approx$ 5.2 feet.

**6.** 513 feet    $\cos(70°) = \dfrac{1,500}{x}$ so $x = \dfrac{1,500}{\cos(70°)} \approx$ 513 feet.

**7.** 83.9 feet    $\tan(40°) = \dfrac{x}{100}$ so $x = 100\tan(40°) \approx$ 83.9 feet.

**8.** 8.1 feet    $\tan(42.6°) = \dfrac{y}{100}$ so $y = 100\tan(42.6°) \approx$ 92.0 feet. 92 feet – 83.9 feet = 8.1 feet is the height of the flagpole.

## CHAPTER 11

# Trigonometry in Other Triangles

## EXERCISE 11.1

**1.** 9.6

$c^2 = a^2 + b^2 - 2ab\cos\angle C$

$c^2 = 14^2 + 18^2 - 2(14)(18)\cos(32°)$

$c^2 = 196 + 324 - 504(0.8480)$

$c^2 = 520 - 427.392 = 92.608$

$c \approx 9.6$

**2.** 32.3

$c^2 = a^2 + b^2 - 2ab\cos\angle C$

$c^2 = 29^2 + 31^2 - 2(29)(31)\cos(65°)$

$c^2 = 841 + 961 - 1,798(0.4226)$

$c^2 = 1,802 - 759.835 = 1,042.165$

$c \approx 32.3$

**3.** 17.6

$c^2 = a^2 + b^2 - 2ab\cos\angle C$

$c^2 = 21^2 + 17^2 - 2(21)(17)\cos(54°)$

$c^2 = 441 + 289 - 714(0.5878)$

$c^2 = 730 - 419.689 = 310.311$

$c \approx 17.6$

**4.** 33.5

$c^2 = a^2 + b^2 - 2ab\cos\angle C$

$c^2 = 33^2 + 44^2 - 2(33)(44)\cos(49°)$

$c^2 = 1,089 + 1,936 - 2,904(0.6561)$

$c^2 = 3,025 - 1,905.314 = 1,119.686$

$c \approx 33.5$

**5.** 82°

$b^2 = a^2 + c^2 - 2ac\cos\angle B$

$22^2 = 14^2 + 19^2 - 2(14)(19)\cos(\angle B)$

$484 = (196 + 361) - 532\cos(\angle B)$

$484 - 557 = -532\cos(\angle B)$

$\cos(\angle B) = \dfrac{484 - 557}{-532} = \dfrac{-73}{-532} \approx 0.1372$

$m\angle B = \cos^{-1}(0.1372) \approx 82°$

**6.** 59°

$a^2 = c^2 + b^2 - 2cb\cos\angle A$

$19^2 = 14^2 + 22^2 - 2(14)(22)\cos(\angle A)$

$361 = (196 + 484) - 616\cos(\angle A)$

$361 - 680 = -616\cos(\angle A)$

$\cos(\angle A) = \dfrac{361 - 680}{-616} = \dfrac{-319}{-616} \approx 0.5179$

$m\angle A = \cos^{-1}(0.5179) \approx 59°$

**7.** 77°

$48^2 = 35^2 + 42^2 - 2(35)(42)\cos(\angle X)$

$2,304 = 1,225 + 1,764 - 2,940\cos(\angle X)$

$2,304 = (1,225 + 1,764) - 2,940\cos(\angle X)$

$2,304 - 2,989 = -2,940\cos(\angle X)$

$\cos(\angle X) = \dfrac{2,304 - 2,989}{-2,940} = \dfrac{-685}{-2,940} \approx 0.2330$

$m\angle X = \cos^{-1}(0.2330) \approx 77°$

## EXERCISE 11.2

**1.** 10.5

$$\frac{15}{\sin 73°} = \frac{x}{\sin 42°}$$

$$x\sin 73° = 15\sin 42°$$

$$x = \frac{15\sin 42°}{\sin 73°} \approx 10.5$$

**2.** 12.6

$$\frac{38}{\sin 100°} = \frac{x}{\sin 19°}$$

$$x = \frac{38\sin 19°}{\sin 100°} \approx 12.6$$

**3.** 108.5

$$\frac{120}{\sin 62°} = \frac{x}{\sin 53°}$$

$$x = \frac{120\sin 53°}{\sin 62°} \approx 108.5$$

**4.** 60.2

$$\frac{85}{\sin 93°} = \frac{x}{\sin 45°}$$

$$x = \frac{85\sin 45°}{\sin 93°} \approx 60.2$$

**5.** No such triangle

$$\frac{22}{\sin 55°} = \frac{34}{\sin(\angle C)}$$

$$22\sin(\angle C) = 34\sin 55°$$

$$\sin(\angle C) = \frac{34\sin 55°}{22} \approx 1.266$$

$$m\angle C = \sin^{-1}(1.266)$$

No solution is possible because sine can only take values between –1 and 1. No such triangle exists.

**6.** Two triangles are possible:

$$\frac{63}{\sin(\angle X)} = \frac{71}{\sin 98°}$$

$$\sin(\angle X) = \frac{63\sin 98°}{71} \approx 0.8787$$

$$m\angle X = \sin^{-1}(0.8787) \approx 62°$$

The triangle may have m∠Z = 98°, m∠X = 62°, and m∠Y = 20°. Also consider ∠Z = 98°, m∠X = 180° – 62° = 118°, but that already exceeds 180°, so a second triangle is not possible.

**7.** Two triangles are possible:

$$\frac{83}{\sin 14°} = \frac{125}{\sin(\angle S)}$$

$$\sin(\angle S) = \frac{125\sin 14°}{83} \approx 0.3643$$

$$m\angle S = \sin^{-1}(0.3643) \approx 21°$$

Consider both 21° and 180° – 21° = 159°. Triangle # 1: 14°, 21°, 145°. Triangle # 2: 14°, 159°, 7°. Both triangles are possible. There are two solutions.

## EXERCISE 11.3

**1.** m∠A = 24°, m∠B = 49°, m∠C = 103°

$$6^2 = 11^2 + 14^2 - 2(11)(14)\cos(\angle A)$$

$$36 = (121 + 196) - 308\cos(\angle A)$$

$$36 - 317 = -308\cos(\angle A)$$

$$\cos(\angle A) = \frac{36 - 317}{-308} = \frac{-281}{-308} \approx 0.9123$$

$$m\angle A = \cos^{-1}(0.9123) \approx 24°$$

$$11^2 = 6^2 + 14^2 - 2(6)(14)\cos(\angle B)$$

$$121 = (36 + 196) - 168\cos(\angle B)$$

$$121 - 232 = -168\cos(\angle B)$$

$$\cos(\angle B) = \frac{121 - 232}{-168} = \frac{-111}{-168} \approx 0.6607$$

$$m\angle B = \cos^{-1}(0.6607) \approx 49°$$

$$m\angle A = 24°, m\angle B = 49°, m\angle C = 103°$$

**2.** $BC \approx 90.1$, $AC \approx 101.9$

m$\angle A = 62°$ and m$\angle B = 88°$ (given) so m$\angle C =$
$180° - (62° + 88°) = 30°$.

$$\frac{51}{\sin 30°} = \frac{BC}{\sin 62°}$$

$$BC = \frac{51 \sin 62°}{\sin 30°} \approx 90.1$$

$$\frac{51}{\sin 30°} = \frac{AC}{\sin 88°}$$

$$AC = \frac{51 \sin 88°}{\sin 30°} \approx 101.9$$

**3.** $XZ = 68.7$, m$\angle Z \approx 42.5°$

$$c^2 = 47^2 + 58^2 - 2(47)(58)\cos(81°)$$
$$c^2 = 2{,}209 + 3{,}364 - 5{,}452(0.1564)$$
$$c^2 = 5{,}573 - 852.6928 = 4{,}720.3072$$
$$c \approx 68.7$$

With side $XZ = 68.7$, find m$\angle Z$.

$$47^2 = 58^2 + 68.7^2 - 2(58)(68.7)\cos(\angle Z)$$
$$2{,}209 = 8{,}083.69 - 7{,}962.2\cos(\angle Z)$$
$$2{,}209 - 8{,}083.69 = -7{,}962.2\cos(\angle Z)$$

$$\cos(\angle Z) = \frac{2{,}209 - 8{,}083.69}{-7{,}962.2} \approx 0.7378$$

m$\angle Z = \cos^{-1}(0.7378) \approx 42.5°$

**4.** $RT = 26.6$ OR $10.0$

$$\frac{19}{\sin 43°} = \frac{25}{\sin(\angle R)}$$

$$\sin(\angle R) = \frac{25 \sin 43°}{19} \approx 0.8974$$

m$\angle R = \sin^{-1}(0.8974) \approx 64°$

Triangle 1: m$\angle R = 64°$, m$\angle S = 180° - (64° +$
$43°) = 73°$, m$\angle T = 43°$

Triangle 2: m$\angle R = 180° - 64° = 116°$, m$\angle S =$
$180° - (116° + 43°) = 21°$, m$\angle T = 43°$

Find side $\overline{RT}$:

$$\frac{RT}{\sin(\angle S)} = \frac{19}{\sin 43°}$$

$$RT = \frac{19 \sin(\angle S)}{\sin 43°} = \frac{19 \sin 73°}{\sin 43°} \text{ OR } \frac{19 \sin 21°}{\sin 43°}$$

$RT = 26.6$ OR $10.0$

**5.** $x \approx 39.5$, $x \approx 42.9$    Let $x =$ length of side
opposite 65° angle and $y =$ length of side
opposite 80°angle.

$$\frac{x}{\sin 65°} = \frac{25}{\sin 35°}$$

$$x = \frac{25 \sin 65°}{\sin 35°} \approx 39.5$$

$$\frac{y}{\sin 80°} = \frac{25}{\sin 35°}$$

$$y = \frac{25 \sin 80°}{\sin 35°} \approx 42.9$$

**6.** m$\angle A = 54°$, m$\angle B = 61°$, m$\angle C = 65°$

$$200^2 = 217^2 + 226^2 - 2(217)(226)\cos \angle A$$
$$40{,}000 = (47{,}089 + 51{,}076) - 98{,}084 \cos \angle A$$
$$40{,}000 - 98{,}165 = -98{,}084 \cos \angle A$$

$$\cos \angle A = \frac{40{,}000 - 98{,}165}{-98{,}084} = \frac{-58{,}165}{-98{,}084} \approx 0.5930$$

m$\angle A = \cos^{-1}(0.5930) \approx 54°$

$$217^2 = 200^2 + 226^2 - 2(200)(226)\cos \angle B$$
$$47{,}089 = (40{,}000 + 51{,}076) - 90{,}400 \cos \angle B$$
$$47{,}089 - 91{,}076 = -90{,}400 \cos \angle B$$

$$\cos \angle B = \frac{-43{,}987}{-90{,}400} \approx 0.4866$$

m$\angle B = \cos^{-1}(0.4866) \approx 61°$

m$\angle A = 54°$, m$\angle B = 61°$, m$\angle C = 180° - (54° +$
$61°) = 65°$

**7.** m$\angle A = 55°$, m$\angle B = 33°$, m$\angle C = 92°$

$$c^2 = 24^2 + 36^2 - 2(24)(36)\cos 92°$$
$$c^2 = 576 + 1{,}296 - 1{,}728(-0.0349)$$
$$c^2 = 1{,}872 + 60.306 = 1{,}932.306$$
$$c \approx 44.0$$

$$\frac{36}{\sin \angle A} = \frac{44.0}{\sin 92°}$$

$$\sin \angle A = \frac{36 \sin 92°}{44.0} = \frac{36(0.9994)}{44.0} = 0.8177$$

m$\angle A = \sin^{-1}(0.8177) \approx 55°$

m$\angle A = 55°$, m$\angle B = 180° - (55° + 92°) = 33°$,
m$\angle C = 92°$

## EXERCISE 11.4

**1.** 101.9

$A = \frac{1}{2}ab\sin(\angle C)$

$A = \frac{1}{2}(15)(18)\sin 49° \approx 101.9$

**2.** 87.0

$A = \frac{1}{2}ab\sin(\angle C)$

$A = \frac{1}{2}(27)(31)\sin 12° \approx 87.0$

**3.** 1,051.9

$A = \frac{1}{2}ab\sin(\angle C)$

$A = \frac{1}{2}(44)(50)\sin 107° \approx 1,051.9$

**4.** 6,990.4

$A = \frac{1}{2}ab\sin(\angle C)$

$A = \frac{1}{2}(112)(125)\sin 93° \approx 6,990.4$

**5.** 65°

$A = \frac{1}{2}ab\sin(\angle C)$

$413.3 = \frac{1}{2}(24)(38)\sin(\angle C)$

$413.3 = 456\sin(\angle C)$

$\sin(\angle C) = \frac{413.3}{456} \approx 0.9064$

$m\angle C = \sin^{-1}(0.9064) \approx 65°$

**6.** 65

$A = \frac{1}{2}ab\sin(\angle C)$

$1,287.3 = \frac{1}{2}(40)(b)\sin 82°$

$1,287.3 = 20(b)(0.9903)$

$b = \frac{1,287.3}{20(0.9903)} \approx \frac{1,287.3}{19.81} \approx 65$

**7.** 24

$A = \frac{1}{2}ab\sin(\angle C)$

$190.7 = \frac{1}{2}(35)(b)\sin 27°$

$190.7 = 7.945b$

$b = \frac{190.7}{7.945} \approx 24$

# REVIEW 5

# Similarity, Right Triangle Trigonometry, and Trigonometry in Other Triangles

**1.** $\triangle ABC \sim \triangle QRP$

**2.** $\frac{WX}{VX} = \frac{XY}{XZ} = \frac{WY}{VZ}$   $\triangle WXY \sim \triangle VXZ$, so

$\frac{WX}{VX} = \frac{XY}{XZ} = \frac{WY}{VZ}$

**3.** $\frac{BC}{CD} = \frac{AC}{CE}$   If $\overline{AB} \parallel \overline{ED}$ (given), then $\angle A \cong \angle E$

and $\angle B \cong \angle D$ because alternate interior angles are congruent. $\triangle ABC \sim \triangle EDC$ by AA, and corresponding sides are in proportion, so

$\frac{BC}{CD} = \frac{AC}{CE}$.

**4.** $\triangle XWZ \sim \triangle WYZ$ by AA   $\angle X \cong \angle ZWY$ (given) and $\angle Z \cong \angle Z$ (reflexive), so $\triangle XWZ \sim \triangle WYZ$ by AA.

**5.** $\frac{BD}{DA} = \frac{BE}{EC}$   Given that $\angle DEA \cong \angle EAC$, it is possible to conclude that $\overline{DE} \parallel \overline{AC}$ by Alternate Interior Angles converse. Then by the Corresponding Angle theorem, $\angle BDE \cong \angle BAC$ and $\angle BED \cong \angle BCA$, so $\triangle BDE \sim \triangle BAC$ by AA. Corresponding sides are in proportion, so

$\frac{BD}{DA} = \frac{BE}{EC}$.

**6.** $\dfrac{PS}{PR} = \dfrac{SQ}{RQ}$   If $\overline{RS} \cong \overline{SQ}$ (given), then $\angle Q \cong$

$\angle SRQ$. Also given is $\angle PRS \cong \angle Q$, and by transitivity, $\angle SRQ \cong \angle PRS$. By the definition of angle bisector, $\overline{RS}$ bisects $\angle PRQ$. Because the bisector of an angle of a triangle divides the opposite side in a way that is proportional to the adjacent sides, $\dfrac{PS}{PR} = \dfrac{SQ}{RQ}$.

**7.** 20.4 cm   $\dfrac{JK}{DE} = \dfrac{JL}{DF}$ becomes $\dfrac{17}{x} = \dfrac{25}{30}$.
Cross-multiply to get $25x = 510$, and solve to get $x = 20.4$ cm.

**8.** 63 cm   $\dfrac{AB}{XY} = \dfrac{BC}{YZ}$ becomes $\dfrac{x-3}{2x+4} = \dfrac{3}{21}$.
Cross-multiply to get $21x - 63 = 6x + 12$, and solve to get $15x = 75$ and $x = 5$. $AC = 5 - 1 = 4$ and $XY = 2(5) + 4 = 14$. $\dfrac{BC}{YZ} = \dfrac{AC}{XZ}$ becomes $\dfrac{3}{21} = \dfrac{4}{y}$. Solve to find $3y = 84$ and $y = 28$. The perimeter of $\triangle XYZ = XY + YZ + XZ = 14 + 21 + 28 = 63$ cm.

**9.** 7.2 feet   Let $x =$ the distance from the point where the man stands to the tip of his shadow.
Solve $\dfrac{10}{18} = \dfrac{6}{x}$ to get $x = 10.8$ feet. To have the tip of the man's shadow match the tip of the tree's shadow, he must stand $18 - 10.8 = 7.2$ feet from the base of the tree.

**10.** 29.75 cm   Convert all measurements to centimeters to avoid confusion if you wish, but you can work with numerators in meters and denominators in centimeters, if you work carefully. $\dfrac{15,300}{38.25} = \dfrac{11,900}{x}$ becomes $15{,}300x = 38.25(11{,}900)$ and $x \approx 29.75$ cm.

**11.** a) $\sin\angle R = \dfrac{33}{65}$

b) $\tan\angle T = \dfrac{56}{33}$

c) $\cos\angle R = \dfrac{56}{65}$

d) $\cos\angle T = \dfrac{33}{65}$

e) $\csc\angle R = \dfrac{1}{\sin\angle R} = \dfrac{65}{33}$

**12.** 21.5 centimeters   $\sin(34°) = \dfrac{12}{h}$ becomes
$h = \dfrac{12}{\sin(34°)} \approx 21.5$ centimeters.

**13.** 36.9°   $\tan(x) = \dfrac{3}{4}$ so $x = \tan^{-1}\left(\dfrac{3}{4}\right) \approx 36.9°$

**14.** 5   $\csc(\angle R) = \dfrac{1}{\sin(\angle R)} = \dfrac{1}{0.2} = 5$

**15.** 5   $\sec(\angle T) = \csc(\angle R) = 5$

**16.** 0.98   $\cos(\angle R) = \sqrt{1 - \sin^2(\angle R)}$
$= \sqrt{1 - 0.04} = \sqrt{0.96} \approx 0.98$

**17.** 19.5 cm   $m\angle J = 63°$, $m\angle K = 180° - (63 + 51) = 66°$, and $m\angle L = 51°$. $JL = 20$ cm. The triangle is not a right triangle, so use the Law of Sines.
$\dfrac{KL}{\sin\angle J} = \dfrac{JL}{\sin\angle K}$ becomes $\dfrac{x}{\sin(63°)} = \dfrac{20}{\sin(66°)}$.
Cross-multiply to get $x\sin(66°) = 20\sin(63°)$ and solve for $x = \dfrac{20\sin(63°)}{\sin(66°)} \approx 19.5$ cm.

**18.** 61.7°
$(YZ)^2 = (XY)^2 + (XZ)^2 - 2(XY)(XZ)\cos(\angle X)$
$(52)^2 = (45)^2 + (55)^2 - 2(45)(55)\cos(\angle X)$
$2{,}704 = (2{,}025 + 3{,}025) - 4{,}950\cos(\angle X)$
$2{,}704 - 5{,}050 = -4{,}950\cos(\angle X)$
$\cos(\angle X) = \dfrac{-2{,}346}{-4{,}950}$
$m\angle X \approx \cos^{-1}(0.4739) \approx 61.7°$

**19.** 195.8 square inches
$A = \tfrac{1}{2}(AB)(BC)\sin(\angle B) = \tfrac{1}{2}(18)(24)\sin(65°)$
$= 216(0.9063) \approx 195.8$ square inches.

**20.** 25°
$A = \tfrac{1}{2}(XY)(YZ)\sin(\angle Y)$
$775.9 = \tfrac{1}{2}(54)(68)\sin(\angle Y)$
$775.9 = 1{,}836\sin(\angle Y)$
$\sin(\angle Y) = \dfrac{775.9}{1{,}836} \approx 0.4226$
$m\angle Y \approx 25°$

## CHAPTER 12

# Circles

## EXERCISE 12.1

1. $\overline{ST}$
2. $\overleftrightarrow{XY}$
3. $\overrightarrow{AB}$
4. $\overset{\frown}{ST}$

5. Radius
6. Common internal tangent
7. Diameter (or chord)

## EXERCISE 12.2

1. **14.1 feet**   Diameter $= 54$ in $= 4.5$ feet, so $C = 4.5\pi \approx 14.1$ feet.

2. $r = 42$ cm $C = 2\pi r = 84\pi$, so $2r = 84$ and $r = 42$ cm.

3. $A = \pi r^2 = 36\pi$ square inches.

4. $r = 17$ meters   $A = 907.92 = \pi r^2$ so $r^2 = 289$ and $r = 17$ meters.

5. **796 square meters**   $C = 100 = 2\pi r$,

   so $r = \dfrac{100}{2\pi} = \dfrac{50}{\pi}$. Then

   $A = \pi r^2 = \cancel{\pi}\left(\dfrac{50}{\cancel{\pi}}\right)\left(\dfrac{50}{\pi}\right) = \dfrac{2,500}{\pi} \approx 796$

   square meters.

6. **10-inch pizza**   A pizza with a diameter of 12 inches has a radius of 6 inches and an area of $36\pi$ square inches. Divide the cost of $12 by $36\pi$ to get $0.1061 per square inch. A pizza with a diameter of 10 inches has a radius of 5 inches and an area of $25\pi$ square inches, so costs $8 \div 25\pi = \$0.1019$ per square inch. The 10-inch pizza is a slightly better buy.

7. $A = \dfrac{C^2}{4\pi}$   Solve $C = 2\pi r$ for $r$:

   $r = \dfrac{C}{2\pi}$. Substitute in the area formula

   $A = \pi r^2 = \cancel{\pi}\left(\dfrac{C}{2\cancel{\pi}}\right)\left(\dfrac{C}{2\pi}\right)$. $A = \dfrac{C^2}{4\pi}$.

## EXERCISE 12.3

1. m$\angle DBC = 19°$   The measure of inscribed angle $\angle DBC$ is half the measure of the central angle with the same arc. m$\angle DBC = \frac{1}{2}$ m$\angle DOC = 19°$.

2. m$\angle PAB = \frac{1}{2}(50°) = 25°$.

3. m$\overset{\frown}{AC} = 42°$   $\angle ABC$ is an inscribed angle, so its measure is half the measure of its intercepted arc; therefore m$\overset{\frown}{AC} = 42°$.

4. m$\angle QAB = 155°$   $\angle QAB$ intercepts $\overset{\frown}{ACB}$. m$\overset{\frown}{ACB} = 360° - 50° = 310°$. m$\angle QAB = \frac{1}{2}(310°) = 155°$.

5. m$\overset{\frown}{BD} = 230°$   m$\overset{\frown}{AB} +$ m$\overset{\frown}{AC} +$ m$\overset{\frown}{CD} = 50° + 42° + 38° = 130°$. m$\overset{\frown}{BD} = 360° - 130° = 230°$.

6. **115°**   m$\angle BAD = \frac{1}{2}(230°) = 115°$.

7. **12.5°**   m$\angle BPD = \frac{1}{2}($m$\overset{\frown}{BD} -$ m$\overset{\frown}{AC}) = \frac{1}{2}(40° - 15°) = \frac{1}{2}(25°) = 12.5°$.

8. **90°**   m$\angle RQT = \frac{1}{2}($m$\overset{\frown}{RBT} -$ m$\overset{\frown}{RT}) = \frac{1}{2}(270° - 90°) = \frac{1}{2}(180°) = 90°$.

9. **135°**   m$\angle SRT = \frac{1}{2}($m$\overset{\frown}{RBT}) = \frac{1}{2}(30° + 40° + 110° + 15° + 75°) = \frac{1}{2}(270°) = 135°$.

**10.** 22.5°  $m\angle DPQ = \frac{1}{2}(m\overset{\frown}{DT} - m\overset{\frown}{CT}) =$
$\frac{1}{2}(120° - 75°) = \frac{1}{2}(45°) = 22.5°.$

**11.** 135°  $m\angle RTP = \frac{1}{2}(m\overset{\frown}{TBR}) = \frac{1}{2}(75° + 15° + 110°$
$+ 40° + 30°) = \frac{1}{2}(270°) = 135°.$

## EXERCISE 12.4

**1.** $EC = 10$   $AE \cdot EB = DE \cdot EC$ becomes $5 \cdot 8 = 4 \cdot EC$, so $EC = 10$.

**2.** $OJ = 8$   $\overline{OK}$ bisects $\overline{FG}$ so FJ = JG = 6.
OK = 10 so OF = 10 (all radii are congruent).
$(OJ)^2 + 6^2 = 10^2$ becomes
$(OJ)^2 = 100 - 36 = 64$, so $OJ = 8$.

**3.** 2  $JK = OK - OJ = 10 - 8 = 2$.

**4.** $OP \approx 21.8$   $(OK)^2 + (OP)^2 = (PK)^2$ becomes
$(OP)^2 = (24)^2 - (10)^2 = 576 - 100 = 476$ so
$OP = \sqrt{476} \approx 21.8$

**5.** $\overline{PD}$ and $\overline{PK}$ are two tangents drawn to the circle from the same point, and so are congruent. Therefore, $\triangle PDK$ is isosceles and by the Base Angle theorem $\angle PDK \cong \angle PKD$.

**6.** $x = 8\frac{1}{3}$   $PA \cdot PB = PC \cdot PD$ becomes $9(9 + 16)$
$= x \cdot 27$ or $27x = 225$, so $x = 8\frac{1}{3}$.

**7.** $PT = 15$   $PC \cdot PD = PT^2$ becomes
$8\frac{1}{3}(27) = PT^2$ and $PT^2 = 225$ and $PT = 15$.

**8.** $QS = 6$   $QR^2 = QS \cdot QV$ becomes $144 = QS \cdot 24$ and $QS = 6$.

**9.** 18  The diameter = $QV - QS = 24 - 6 = 18$.

**10.** 17  If $x(x + 1) = 56$, solve $x^2 + x - 56 = 0$ by factoring. Setting each factor in $(x + 8)(x - 7) = 0$ equal to zero will produce solutions of $x = -8$, which is rejected, and $x = 7$. Then $AB = x + (x + 1) = 7 + 8 = 15$, and $CD = (x + 1) + (x + 2) = 8 + 9 = 17$.

## EXERCISE 12.5

**1.** Draw a radius from $C$ to $A$ and extend. Scribe a segment which has $A$ as its midpoint. Construct the perpendicular bisector.

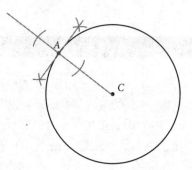

Draw tangents to the two points where the arc intersects the circle.

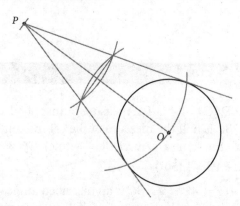

**2.** Draw $\overline{OP}$ and bisect it. Scribe an arc centered at the midpoint of $\overline{OP}$ with a radius that is $\frac{1}{2}OP$.

**3.** Draw a radius. From a point on the circle, with the compass set to the radius of the circle, scribe arcs that divide the circle into six congruent arcs. Connect the points.

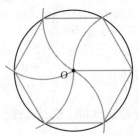

**5.**

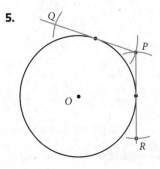

**4.** Draw $\overline{OP}$ and bisect it. With the midpoint as center and $\frac{1}{2}OP$ as radius, scribe a circle that intersects the original circle in two points. Draw from $P$ to each point of intersection.

**6.**

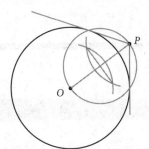

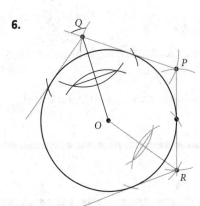

**7.** Draw a diameter then construct a perpendicular to it, through the center, to divide the circle into four congruent segments. Construct a perpendicular at a point where the diameter crosses the circle. Repeat at each point.

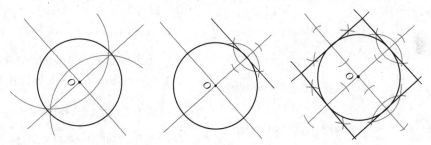

# EXERCISE 12.6

**1.** 67 meters  Length of $\widehat{AB} = \dfrac{40}{360} \cdot 2\pi \cdot 12 = \dfrac{24\pi}{9} = \dfrac{8\pi}{3} \approx 8.4$ meters, and length of

$$\widehat{ADB} = C - \dfrac{8\pi}{3} = 24\pi - \dfrac{8\pi}{3} = \dfrac{64\pi}{3} \approx 67 \text{ meters.}$$

**2.** $48\pi$ square meters  $A = \dfrac{120}{360}\pi r^2 = \dfrac{120}{360}(144\pi) = 48\pi$ square meters.

**3.** 49.1 meters  The perimeter of the sector is the combined lengths of two radii and the arc.

$2 \cdot OA + \text{length of } \widehat{DG} = 2(12) + \frac{1}{3}(24\pi) = 24 + 8\pi \approx 49.1 \text{ meters}$

**4.** 29.8 cm    $s = \dfrac{57}{360} \cdot 2\pi \cdot 30 \approx 29.8$ cm

**5.** 13.2 square inches    If $d = 12$, $r = 6$.

$A = \dfrac{42}{360} \cdot \pi \cdot 36 = 4.2\pi \approx 13.2$ square inches

**6.** The first arc is longer.    The first arc has a length of $\dfrac{75}{360} \cdot 2\pi \cdot 80 = \dfrac{100\pi}{3}$ cm, and the second arc has a length of $\dfrac{80}{360} \cdot 2\pi \cdot 72 = 32\pi$ cm. The first arc is longer.

**7.** The first sector has an area twice as large as the second. The first sector has an area of

$\dfrac{30}{360} \cdot \pi \cdot (24)^2 = 48\pi$ square centimeters, and

the second sector has an area of $\dfrac{60}{360} \cdot \pi \cdot (12)^2 =$ $24\pi$ square centimeters. The first sector has an area twice as large as the second.

**8.** 22.3 square centimeters    The sector with the larger radius has an area of $\dfrac{40}{360} \cdot \pi \cdot 10^2 = \dfrac{100\pi}{9}$. With the smaller radius, the area is $\dfrac{40}{360} \cdot \pi \cdot 6^2 =$ $\dfrac{36\pi}{9}$. The shaded area is $\dfrac{100\pi}{9} - \dfrac{36\pi}{9} =$ square centimeters.

## EXERCISE 12.7

**1.** $\dfrac{\pi}{6}$ radians    $\dfrac{30}{360} = \dfrac{r}{2\pi}$ becomes $360r = 60\pi$ and $r = \dfrac{\pi}{6}$ radians. (approximately 0.524 radians)

**2.** $\dfrac{3\pi}{4}$ radians    $\dfrac{135}{360} = \dfrac{r}{2\pi}$ becomes $360r = 270\pi$ and $r = \dfrac{3\pi}{4}$ radians. (approximately 2.356 radians)

**3.** $\dfrac{4\pi}{9}$ radians    $\dfrac{80}{360} = \dfrac{r}{2\pi}$ becomes $360r = 160\pi$ and $r = \dfrac{160\pi}{360} = \dfrac{4\pi}{9}$ radians. (approximately 1.396 radians)

**4.** 60°    $\dfrac{d}{360} = \dfrac{\pi/3}{2\pi}$ becomes $2\pi d = 120\pi$ and $d = 60°$.

**5.** 180°    $\dfrac{d}{360} = \dfrac{\pi}{2\pi}$ becomes $2\pi d = 360\pi$ and $d = 180°$.

**6.** 143°    $\dfrac{d}{360} = \dfrac{2.5}{2\pi}$ becomes $2\pi d = 900$ and $d \approx 143°$.

**7.** 34°    $\dfrac{d}{360} = \dfrac{0.6}{2\pi}$ becomes $2\pi d = 216$ and $d \approx 34°$.

**8.** 62.8 inches    $s = \theta r$ becomes $s = \dfrac{5\pi}{6} \cdot 24 = 20\pi$ inches, approximately 62.8 inches.

**9.** 1.5 radians    $s = \theta r$ becomes $108 = \theta \cdot 72$, so $\theta = \dfrac{108}{72} = 1.5$ radians.

**10.** 19.6 square meters    $A = \dfrac{\theta}{2} r^2$ becomes $A = \dfrac{\pi/2 \cdot 5^2}{2} = \dfrac{25\pi}{2} \div 2 = \dfrac{25\pi}{4}$ square meters, or approximately 19.6 square meters.

## CHAPTER 13

# Conic Sections

## EXERCISE 13.1

**1.** $C(3, 2), r = 2$

**2.** $C(-3, -5), r = 11$

**3.** $x^2 + y^2 = 49$

**4.** $(x - 2)^2 + (y - 9)^2 = 100$

**5.** $(x + 7)^2 + (y + 4)^2 = 64$

**6.**

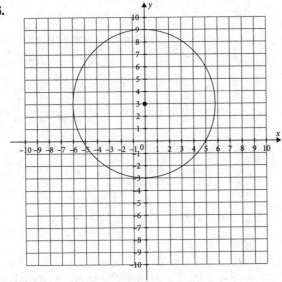

**7.**

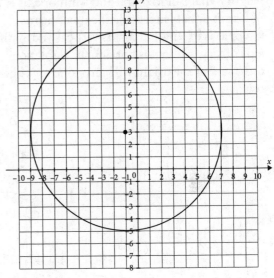

## EXERCISE 13.2

**1.** 10 units    The major axis is $2a$ and $a = \sqrt{25} = 5$, so the major axis is 10 units long.

**2.** 14 units    The minor axis is $2b$ and $b = 7$, so the minor axis is 14 units.

**3.** $\dfrac{x^2}{36} + \dfrac{y^2}{81} = 1$    $C(0, 0)$, minor axis (horizontal) $= 12$ so $b = 6$, major axis (vertical) $= 18$, so $a = 9$. $\dfrac{x^2}{36} + \dfrac{y^2}{81} = 1$

**4.** $\dfrac{(x+3)^2}{25} + \dfrac{(y+2)^2}{9} = 1$  $C(-3, -2)$, major

axis (horizontal) = 10 so $a = 5$, minor axis

(vertical) = 6 so $b = 3$. $\dfrac{(x+3)^2}{25} + \dfrac{(y+2)^2}{9} = 1$

**5.** A rotation of 90° about the origin will carry one onto the other. Both are ellipses centered at the origin. One has a major axis of length 8 that is horizontal and a minor axis of length 6 that is vertical. The other has a major axis of length 8 that is vertical and a minor axis of length 6 that is horizontal. A rotation of 90° about the origin will carry one onto the other.

**7.**

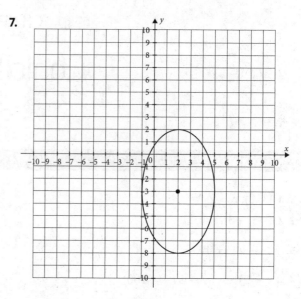

**6.**

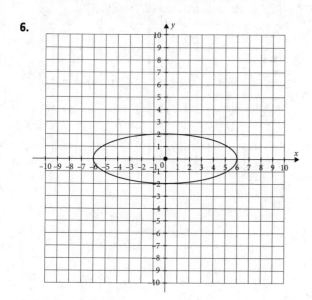

**8.**

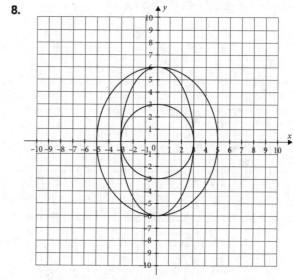

**9.** $6\pi$, $10\pi$  The area of the ellipse is $A = \pi ab = \pi \cdot 3 \cdot 5 = 15\pi$. The area of the smaller circle is $A = \pi r^2 = 9\pi$. The difference is $6\pi$. The area of the larger circle is $A = \pi r^2 = 25\pi$, so the difference between the area of the larger circle and the area of the ellipse is $10\pi$.

## EXERCISE 13.3

**1.** a) down

  b) right

  c) up

**2.** a) $(-2, 6)$

  b) $(-4, 3)$

  c) $(-5, 4)$

**3.** $y = -2x^2 + 5$  The parabola opens down, has a vertex of $(0, 5)$, and passes through the point $(1, 3)$. Begin with $y - k = a(x - h)^2$ and replace $h$ with 0 and $k$ with 5. $y - 5 = a(x - 0)^2$ or $y - 5 = ax^2$. Then replace $x$ with 1 and $y$ with 3 to solve for $a$. $3 - 5 = a(1)^2$ says $a = -2$. The equation is $y - 5 = -2x^2$ or $y = -2x^2 + 5$.

**4.** $x = \frac{1}{2}(y + 2)^2 - 6$   The parabola opens
to the right, has a vertex of $(-6, -2)$, and
passes through the point $(-4, 0)$. Begin with
$x - h = a(y - k)^2$ and substitute $-6$ for $h$
and $-2$ for $k$. With $x + 6 = a(y + 2)^2$,
substitute $-4$ for $x$ and $0$ for $y$, and solve for $a$.
$-4 + 6 = a(0 + 2)^2$ becomes $2 = 4a$,
so $a = \frac{1}{2}$. The equation is $x + 6 = \frac{1}{2}(y + 2)^2$ or
$x = \frac{1}{2}(y + 2)^2 - 6$.

**5.**

**6.**

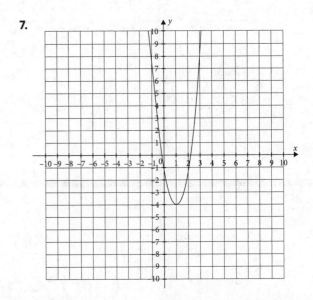

**7.**

---

## EXERCISE 13.4

**1.** a) $(0, 3)$

   b) $(-2, 3)$

   c) $(1, -1)$

**2.** a) left/right

   b) up/down

   c) left/right

**3.** $\dfrac{x^2}{4} - \dfrac{y^2}{9} = 1$   The hyperbola is centered

at the origin, opens to the left and right, and
touches the edges of the rectangle at $(\pm 2, 0)$.

The rectangle is 4 units wide and 6 units tall,

so $b = 2$ and $a = 3$. $\dfrac{x^2}{4} - \dfrac{y^2}{9} = 1$

**4.** $\dfrac{y^2}{1} - \dfrac{x^2}{100} = 1$   The hyperbola is centered

at the origin, opens up and down, and touches
the edges of the rectangle at $(0, \pm 1)$. The rectangle
is 20 units wide and 2 units tall, so $a = 10$ and
$b = 1$.

$\dfrac{y^2}{1} - \dfrac{x^2}{100} = 1$

**5.** $\dfrac{(x-2)^2}{25} - \dfrac{(y+1)^2}{36} = 1$   The hyperbola is centered at $(2, -1)$, and opens to the left and right. The rectangle is 10 units wide and 12 units tall, so $b = 5$ and $a = 6$.

$$\dfrac{(x-2)^2}{25} - \dfrac{(y+1)^2}{36} = 1$$

**7.**

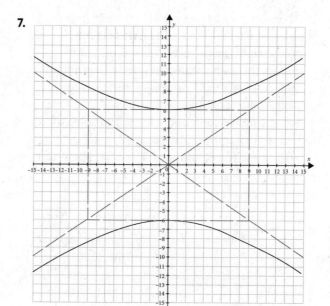

**6.**

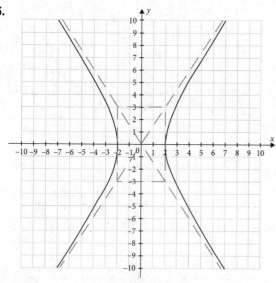

---

## REVIEW 6

# Circles and Conics

**1.** $\overline{ST}$

**2.** $\angle RST$

**3.** $m\angle RST = \frac{1}{2}m\overset{\frown}{RT}$

**4.** $r = 12$

**5.** $\frac{3}{4}$ of the large circle is shaded   Let $D =$ diameter of the larger circle, $R =$ radius of the larger circle, $d =$ diameter of the smaller circle, and $r =$ the radius of the smaller circle. $r = \frac{1}{2}R = \frac{1}{2}\left(\frac{1}{2}D\right) = \frac{1}{4}D$. Area of the larger circle: $A_L = \pi\left(\frac{1}{2}D\right)^2 = \frac{1}{4}\pi D^2$. Area of the smaller circle: $A_S = \pi r^2 = \pi\left(\frac{1}{4}D\right)^2 = \frac{1}{16}\pi D^2$. The shaded area is $A_L - A_S = \frac{1}{4}\pi D^2 - \frac{1}{16}\pi D^2 = \frac{3}{16}\pi D^2$.

This is what part of $A_L = \frac{1}{4}\pi D^2$? $\frac{3}{16} \div \frac{1}{4} = \frac{3}{4}$ of the large circle is shaded.

**6.** $2\pi$ centimeters   $\angle OAE$ is an inscribed angle in circle $C$ that measures $30°$ and so arc $\overset{\frown}{OB}$ measures $60°$. $\angle OAE$ is also an inscribed angle in circle $O$, and so arc $\overset{\frown}{DC}$ also measures $60°$. The central angles that intercept these arcs will measure $60°$. The length of arc $\overset{\frown}{OB}$ is $\pi$ cm, so set up $\dfrac{60°}{360°} \cdot 2\pi r = \pi$ and solve for $r$. $\frac{1}{3}\pi r = \pi$, so $r = 3$ cm and $R = 6$ cm. The length of arc $\overset{\frown}{DC}$ is $\dfrac{60°}{360°} \cdot 2\pi R = \frac{1}{3} \cdot \pi \cdot 6 = 2\pi$ centimeters.

**7.** $m\angle PQR = m\angle PRQ = 80°$  Two tangents drawn from the same point will be congruent, so $\triangle QPR$ is isosceles and $\angle PQR \cong \angle PRQ$. $m\angle P = 20°$, so there is $180° - 20° = 160°$ for the two base angles. $m\angle PQR = m\angle PRQ = 80°$.

**8.** $CD = 14$   $AE \cdot EB = CE \cdot ED$ becomes $x(3x) = 2x(x + 2)$ or $3x^2 = 2x^2 + 4x$. Simplify to $x^2 - 4x = 0$ and solve by factoring to get $x = 0$, which is rejected, and $x = 4$. $CD = 2x + (x + 2) = 8 + 6 = 14$.

**9.** $m\angle AED = 112°$   $m\angle AED = \frac{1}{2}\left(m\widehat{AD} + m\widehat{BC}\right)$ $= \frac{1}{2}(72° + 64°) = \frac{1}{2}(136°) = 68°$. $m\angle AEC = 180° - m\angle AED = 180° - 68° = 112°$

**10.** $m\angle ACB = \frac{1}{2}m\widehat{AB} = \frac{1}{2}(48°) = 24°$

**11.** $x = 7$   $PB \cdot AP = PC \cdot PD$ becomes $3(17 + 3) = 5(5 + x)$ or $60 = 25 + 5x$. Solve to find $x = 7$.

**12.** Draw $\overline{OP}$ and bisect it. From the midpoint of $\overline{OP}$, scribe a circle that passes through $O$ and $P$ and intersects the circle in two points. Those two points are the points of tangency. Draw from $P$ to each point of tangency.

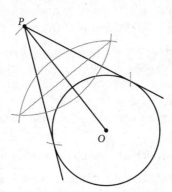

**13.** $m\widehat{BDF} = 240°$   To inscribe a regular hexagon in a circle, the circle is divided into 6 congruent arcs of 60° each. $m\widehat{BDF} = m\widehat{BC} + m\widehat{CD} + m\widehat{DE} + m\widehat{EF} = 4(60°) = 240°$

**14.** $m\angle BPF = \frac{1}{2}(m\widehat{BDF} - m\widehat{FB}) = \frac{1}{2}(240° - 120°) = 60°$

**15.** The length of $\widehat{AB} \approx 10$ inches. The area of the sector is 34 square inches. $A = \frac{85}{360}\pi r^2 = 34$ so $r^2 = 34 \cdot \frac{360}{85\pi} \approx 45.84$. The radius $r \approx \sqrt{45.84} \approx 6.8$ inches. The length of $\widehat{AB}$ is $\frac{85}{360} \cdot 2\pi(6.8) \approx 10$ inches.

**16.** $r = \frac{2\pi}{3}$ radians   $\frac{r}{2\pi} = \frac{120°}{360°}$ cross-multiplies to $360r = 240\pi$ and $r = \frac{240\pi}{360} = \frac{2\pi}{3}$ radians.

**17.** $d = 135°$   $\frac{3\pi/4}{2\pi} = \frac{d}{360°}$ cross-multiplies to $2\pi d = 270\pi$ and $d = 135°$.

**18.**

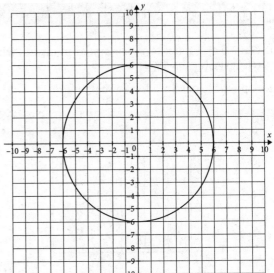

**19.** $(x - 4)^2 + (y - 3)^2 = 25$   The center is $C(4, 3)$ and the radius is $r = 5$. The equation of the circle is $(x - 4)^2 + (y - 3)^2 = 25$.

**20.** a) hyperbola
    b) circle
    c) ellipse
    d) parabola

CHAPTER 14

# Three-Dimensional Figures

## EXERCISE 14.1

1. Triangular pyramid
2. Pentahedron
3. Heptahedron
4. $E = 18$   $F + V = E + 2$ becomes $8 + 12 = E + 2$, so $E = 18$.

5. $F = 4$   $F + V = E + 2$ becomes $F + 4 = 6 + 2$, so $F = 4$.
6. $V = 8$   $F + V = E + 2$ becomes $6 + V = 12 + 2$, so $V = 8$.
7. Hexagonal prisms have 8 faces, 18 edges, and 12 vertices.

## EXERCISE 14.2

1. Choice a) is a net for a triangular pyramid, which has 4 triangular faces.
2. A possible net for a hexagonal prism:

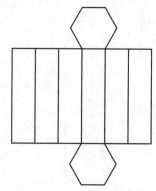

3. 555 square centimeters   $S = 2(\frac{1}{2}aP) + Ph = 2.75(20) + 20(25) = 55 + 500 = 555$ square centimeters.

4. $720\pi$ square centimeters $S = 2\pi r^2 + 2\pi rh = 2\pi(144) + 2\pi(12)(18) = 288\pi + 432\pi = 720\pi$ square inches.

5. $h \approx 16$ cm   $S = 2B + Ph$ becomes $1{,}100.55 = 2(166.28) + 48h$. Solve to find $48h = 1{,}100.55 - 332.56 = 767.99$ and $h \approx 16$ cm.

6. $r = 4$ inches   $S = 2\pi r^2 + 2\pi rh$ becomes $104\pi = 2\pi r^2 + 2\pi r(9)$, which simplifies to $52 = r^2 + 9r$. Solve by factoring. $r^2 + 9r - 52 = (r + 13)(r - 4) = 0$ has solutions of $r = -13$, which is rejected, and $r = 4$ inches.

7. The cylinder has the larger surface area

   If the radius, from center to vertex, of the square, is $r$, then the side of the square is $r\sqrt{2}$ and the area of each square base is $\left(r\sqrt{2}\right)^2 = 2r^2$. The surface area of the square prism is $S = 2B + Ph = 2(2r^2) + 4r\sqrt{2}h = 4r^2 + 4\sqrt{2}rh$. The surface area of the cylinder is $S = 2\pi r^2 + 2\pi rh$. Comparing $4r^2 + 4\sqrt{2}rh$ with $2\pi r^2 + 2\pi rh$ means comparing 4 with $2\pi$ and $4\sqrt{2}$ with $2\pi$. Since $4 < 2\pi$ and $4\sqrt{2} < 2\pi$, the cylinder has the larger surface area.

## EXERCISE 14.3

**1.** 2,494.2 cubic inches   If the side of the hexagonal base is 8, then the apothem is $4\sqrt{3}$ and the area of the base is $\frac{1}{2}aP = \frac{1}{2}\left(4\sqrt{3}\right)(48) = 96\sqrt{3}$. Then the volume is $V = Bh = \left(96\sqrt{3}\right)(15) = 1{,}440\sqrt{3}$ cubic inches (approximately 2,494.2 cubic inches).

**2.** 3,769.9 cubic centimeters   If the diameter is 20, then $r = 10$ and $V = \pi r^2 h = \pi(100)(12) = 1{,}200\pi \approx 3{,}769.9$ cubic centimeters.

**3.** $h = 24$   $V = Bh$, so solve $25{,}807.2 = 1{,}075.3h$ to find $h = 24$.

**4.** $h \approx 32$ inches   $V = \pi r^2 h$ becomes $787{,}320 = \pi(81)^2 h$. Solve to find $h \approx 32$ inches.

**5.** $4{,}608\pi$ cubic inches   $S = 2\pi r^2 + 2\pi rh$, with a diameter of 24 and therefore $r = 12$ inches, becomes $1{,}056\pi = 2\pi(12)^2 + 2\pi(12)h = 288\pi + 24\pi h$. Simplify to $44 = 12 + h$ and $h = 32$ inches. Then $V = \pi r^2 h = \pi(12)^2(32) = 4{,}608\pi$ cubic inches.

**6.** The ratio is 5.   The 3-4-5 base has an area of 6, and the 5-12-13 base has an area of 30. If the volumes are to be equal $6h_1 = 30h_2$ and $\dfrac{h_1}{h_2} = \dfrac{30}{6} = 5$.

**7.** The height must be divided by 4.   If the radius is doubled, the volume $V = \pi(2r)^2 h = 4\pi r^2 h$ is multiplied by . If the volume is going to remain the same, the height must be divided by 4.

## EXERCISE 14.4

**1.** 1,036.75 square centimeters   $S = B + \frac{1}{2}Pl = 247.75 + \frac{1}{2}(60)(26.3) = 247.75 + 789 = 1{,}036.75$ square centimeters.

**2.** 933.1 square inches   $S = \pi r^2 + \pi rl = \pi(9)^2 + \pi(9)(24) = 81\pi + 216\pi = 297\pi$ square inches (or approximately 933.1 square inches.)

**3.** 1,433.1 square inches   The side of the hexagon is $96 \div 6 = 16$ inches, and the apothem of the hexagon is $8\sqrt{3}$. The area of the hexagonal base is $B = \frac{1}{2}aP = \frac{1}{2}\left(8\sqrt{3}\right)(96) = 384\sqrt{3}$ square inches. The slant height is 16 inches. The surface area is $S = B + \frac{1}{2}lP = 384\sqrt{3} + \frac{1}{2}(16)(96) \approx 1{,}433.1$ square inches.

**4.** 18.5 inches   $S = B + \frac{1}{2}lP$ becomes $1{,}024.5 = 377 + \frac{1}{2}(70)l$. Solve to find $35l = 647.5$ and $l = 18.5$ inches.

**5.** The second cone has the larger surface area.   A cone with a radius of 3 and a height of 4 has a surface area of $S = \pi r^2 + \pi rl = \pi(3)^2 + \pi(3)(5) = 24\pi$ square centimeters. A cone with a radius of 4 and a height of 3 has a surface area of $S = \pi r^2 + \pi rl = \pi(4)^2 + \pi(4)(5) = 36\pi$ square centimeters. The second cone has the larger surface area.

**6.** 5.4 square feet larger   The shorter pyramid has a slant height of $\sqrt{10}$ and the taller pyramid has a slant height of $\sqrt{13}$. The surface area of the shorter pyramid is $S = 36 + \frac{1}{2}(24)\sqrt{10} \approx 73.9$ square feet. The surface area of the taller pyramid is $S = 36 + \frac{1}{2}(24)\sqrt{13} \approx 79.3$ square feet. The taller pyramid has a surface area $79.3 - 73.9 = 5.4$ square feet larger.

**7.** The cone has the larger surface area.   The square pyramid has a perimeter of 40, a side of 10, an apothem of 5 and a slant height of $\sqrt{425}$. Its surface area is $S = B + \frac{1}{2}Pl = 100 + 20\sqrt{425} \approx 512.3$ square centimeters. The cone, with a slant height of $\sqrt{500}$, has a surface area of $S = \pi r^2 + \pi rl = 100\pi + 10\pi\sqrt{500} \approx 1{,}016.6$ square centimeters. The cone has the larger surface area.

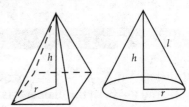

## EXERCISE 14.5

1. 36 square centimeters $V = \frac{1}{3}Bh = \frac{1}{3}(\frac{1}{2} \cdot 3 \cdot 4) \cdot 18 = 36$ square centimeters.

2. 804.2 square inches $V = \frac{1}{3}\pi r^2 h = \frac{1}{3} \cdot \pi \cdot 8^2 \cdot 12 = 256\pi \approx 804.2$ square inches.

3. $r = 6$ cm $V = \frac{1}{3}\pi r^2 h$ becomes $300\pi = \frac{1}{3}\pi r^2(25)$. Solve to find that $r^2 = 36$ and $r = 6$ cm.

4. $h = 11$ cm $V = \frac{1}{3}Bh$ becomes $154 = \frac{1}{3}(42)h$. Solve to find $h = 11$ cm.

5. 37.7 cubic centimeters If then slant height is 5 cm, and the cone has a surface area of $24\pi$ square centimeters then $S = \pi r^2 + \pi r l = \pi r^2 + \pi r(5) = 24\pi$ and $r^2 + 5r = 24$. Solve $r^2 + 5r - 24 = 0$ by factoring or quadratic formula, and $r = -8$ (reject) or $r = 3$. With a radius of 3 and a slant height of 5, use the Pythagorean theorem to find $h = 4$. A cone with a radius of 3 and a height of 4 has a volume of $V = \frac{1}{3}\pi r^2 h = \frac{1}{3}\pi(3^2)(4) = 12\pi$ cubic centimeters, or approximately 37.7 cubic centimeters.

6. 531.7 square centimeters $V = \frac{1}{3}BH$ becomes $720 = \frac{1}{3}B(15)$. Solve to find $B = 144$ square centimeters. The square base has a side of 12 and a perimeter of 48. The surface area of the pyramid is $S = B + \frac{1}{2}Pl = 144 + \frac{1}{2}(48)\sqrt{261} \approx 531.7$ square centimeters.

7. $r = 5$ inches The volume of the cone is $\frac{1}{3}$ the volume of the cylinder, so the empty space is $\frac{2}{3}$ of the volume of the cylinder. If $300\pi = \frac{2}{3}\pi r^2 h$ then $\pi r^2 h = 450\pi$. If $h = 18$, $\pi r^2(18) = 450\pi$ becomes $r^2 = 25$ and $r = 5$ inches.

## EXERCISE 14.6

1. 2,463 square centimeters $S = 4\pi r^2 = 4\pi(14)^2 = 784\pi \approx 2,463$ square centimeters.

2. 268.1 cubic inches $V = \frac{4}{3}\pi r^3 = \frac{4}{3}\pi(4)^3 = \frac{256\pi}{3} \approx 268.1$ cubic inches.

3. $r = 6$ cm $V = \frac{4}{3}\pi r^3$ becomes $288\pi = \frac{4}{3}\pi r^3$. Solve to find $r^3 = 216$ and $r = 6$ cm.

4. 18 inches $S = 4\pi r^2 = 324\pi$. Solve to find $r^2 = 81$ and $r = 9$. The diameter is 18 inches.

5. 1,809.6 square meters $V = \frac{4}{3}\pi r^3 = 2,304\pi$. Solve to find $r^3 = 1,728$ and $r = 12$ meters. Then $S = 4\pi r^2 = 4\pi(144) = 576\pi \approx 1,809.6$ square meters.

6. $r = 3$ cm If $4\pi r^2 = \frac{4}{3}\pi r^3$, then $r^2 = \frac{1}{3}r^3$ so $3r^2 = r^3$ and $r^3 - 3r^2 = r^2(r - 3) = 0$. The equation has two solutions, $r = 0$, which is impossible, and $r = 3$ cm.

7. 5,575.3 cubic centimeters If the diameter is 22 cm, the radius is 11 cm. The volume is $V = \frac{4}{3}\pi r^3 = \frac{4}{3}\pi(11)^3 = \frac{4}{3}\pi(1,331) \approx 5,575.3$ cubic centimeters.

## EXERCISE 14.7

1. 6 centimeters If $V = \frac{4}{3}\pi r^3 = 36\pi$, then $r^3 = 27$ and $r = 3$, so the diameter is 6 cm.

2. $h = 12$ cm $V = \frac{1}{3}\pi r^2 h$ becomes $36\pi = \frac{1}{3}\pi(3)^2 h$. Solve for $h = 12$ cm.

3. 12 feet $10,000$ gallons $\div 7.48$ gallons per cubic foot $\approx 1,336.9$ cubic feet. A cylindrical tank with a height of 12 feet needs to have a volume of 1,336.9 cubic feet. $V = \pi r^2 h$ becomes $1,336.9 = \pi r^2(12)$. Solve to find $r^2 \approx 35.5$ so $r \approx 5.95$. The radius of the tank should be about 6 feet and the diameter approximately 12 feet.

**4.** The radius of 3 cm gives a smaller surface area for the same volume. If the volume of the cylinder is 250 cubic centimeters and the radius is 3 cm, solving $250 = \pi(3)^2 h$ says the height must be approximately 8.84 cm. Those dimensions would result in a surface area of $S = 2\pi r^2 + 2\pi rh = 2\pi(3)^2 + 2\pi(3)(8.84) \approx 71.1\pi$ square centimeters. If the same volume is desired with a radius of 4 cm, $250 = \pi(4)^2 h$ gives $h \approx 5.0$ cm, and yields a surface area of $S = 2\pi r^2 + 2\pi rh = 2\pi(4)^2 + 2\pi(4)(5.0) \approx 72\pi$. The radius of 3 cm gives a smaller surface area for the same volume.

**5.** 180 billion square miles   If the radius of the Earth is 3,959 miles, the surface area will be $S = \frac{4}{3}\pi r^3 = \frac{4}{3}\pi(6.2 \times 10^{10}) \approx 2.6 \times 10^{11}$ square miles. If 71% of that surface is water, that equals $0.71(2.6 \times 10^{11}) \approx 1.8 \times 10^{11}$ square miles. (180 billion square miles)

**6.** The pyramid in Mexico has the larger volume. The pyramid in Egypt has a volume of $V = \frac{1}{3}Bh = \frac{1}{3}(230)^2(146.5) \approx 2{,}583{,}283.3$ cubic meters. The pyramid in Mexico has a volume of $V = \frac{1}{3}Bh = \frac{1}{3}(400)^2(55) \approx 2{,}933{,}333.3$ cubic meters. The pyramid in Mexico has the larger volume.

**7.** The height of the cone must be twice the radius. The cylindrical portion of the volume can be ignored because it is the same in both structures, but the volume of the hemisphere must equal the volume of the cone. $\frac{1}{2}\left(\frac{4}{3}\pi r^3\right) = \frac{1}{3}\pi r^2 h$. The radius is determined by the radius of the cylinder and is the same in both structures. $\frac{2}{3}\pi r^3 = \frac{1}{3}\pi r^2 h$ simplifies to $2r^3 = r^2 h$ or $2r^3 - r^2 h = r^2(2r - h) = 0$. A solution of $r = 0$ must be rejected, but $2r - h = 0$ means $h = 2r$. The height of the cone must be twice the radius, or put another way, the height of the cone is equal to its diameter.

## EXERCISE 14.8

**1.** 3,720.6 grams   The volume of the cylinder is $V = \pi r^2 h = \pi(5.5)^2(14.5) \approx 1{,}378$ cubic centimeters. If cast from aluminum, it will have a mass of $2.7(1{,}378) = 3{,}720.6$ grams.

**2.** An increase of 7,372.3 grams   If the cylinder in exercise 50 is instead cast from steel, its mass will be $8.05(1{,}378) = 11{,}092.9$ grams. This is an increase of 7,372.3 grams.

**3.** 1.11 grams per cubic centimeter   If 355 cubic centimeters of a cola have a mass of 394 grams, the density of the cola is $\frac{394}{355} \approx 1.11$ grams per cubic centimeter.

**4.** 1.0003 grams per cubic centimeter   If 355 cubic centimeters of diet cola has a mass of 355.1 grams, the density of diet cola is $\frac{355.1}{355} \approx 1.0003$ grams per cubic centimeter.

**5.** The key difference between cola and diet cola is the sweetening agent used. The artificial sweeteners used in diet cola are less dense than the sugar used to sweeten cola.

**6.** The difference is 4.556 grams.   The volume of the bead is the volume of a sphere of diameter $= 1$ cm minus the volume of a tiny cylinder with diameter $= 0.1$ cm and height equal to the diameter of the sphere. The radius of the sphere is 0.5 cm, and the radius of the cylinder is 0.05 cm, and the height of the cylinder is 1.

$$V = \frac{4}{3}\pi(0.5)^3 - \pi(0.05)^2(1) = \frac{\pi}{6} - \frac{\pi}{400} \approx 0.516$$

cubic centimeters. If cast in gold, the bead will have a mass of $19.32(0.516) = 9.969$ grams, but if cast in silver, the mass is $10.49(0.516) = 5.413$ grams. The difference is 4.556 grams.

## REVIEW 7

# Three-Dimensional Figures

**1.** $F = 8$   $F + V = E + 2$ becomes $F + 12 = 18 + 2$. Solve to find $F = 8$.

**2.** A possible map for a triangular pyramid:

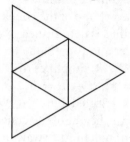

**3.** 56.8 square inches   Perimeter $= 12$ means each side of the hexagon is 2 and the apothem is $\sqrt{3}$. $B = \frac{1}{2}aP = \frac{1}{2}(\sqrt{3})(12) = 6\sqrt{3}$ , and $S = 2B + Ph = 2(6\sqrt{3}) + 12(3) = 12\sqrt{3} + 36 \approx 56.8$ square inches.

**4.** 1,105.8 square centimeters   $S = 2\pi r^2 + 2\pi rh = 2\pi(8)^2 + 2\pi(8)(14) = 128\pi + 224\pi = 352\pi \approx$ 1,105.8 square centimeters.

**5.** 1,320 cubic feet   $B = \frac{1}{2}aP = \frac{1}{2}(5.5)(40) = 110$ square feet. $V = Bh = (110)(12) = 1,320$ cubic feet.

**6.** 18,472.6 cubic inches   $V = \pi r^2 h = \pi(14)^2(30) = 5,880\pi \approx 18,472.6$  cubic inches.

**7.** 85 inches   $l = \sqrt{a^2 + h^2} = \sqrt{36^2 + 77^2} = \sqrt{7,225} \approx 85$ inches.

**8.** 2,358 square centimeters   The area of the base octagon is $B = \frac{1}{2}aP = \frac{1}{2}(18.1)(120) = 1,086$ square centimeters. The slant height is $l = \sqrt{a^2 + h^2} = \sqrt{18.1^2 + 11^2} = \sqrt{448.61} \approx 21.2$ cm, and the surface area is $S = B + \frac{1}{2}lP = 1,086 + \frac{1}{2}(21.2)(120) = 1,086 + 1,272 = 2,358$ square centimeters.

**9.** 678.6 square inches   If diameter $= 18$, then radius $= 9$, height $= 12$ and slant height $= 15$ inches. $S = \pi r^2 + \pi rl = \pi \cdot 9^2 + \pi \cdot 9 \cdot 15 = 81\pi + 135\pi = 216\pi \approx 678.6$ square inches.

**10.** 32 cubic feet   $V = \frac{1}{3}Bh = \frac{1}{3}(16)(6) = 32$ cubic feet.

**11.** 7,938.8 cubic inches   If the circumference is $38\pi$, then the diameter is 38 and the radius is 19 inches. $V = \frac{1}{3}\pi r^2 h = \frac{1}{3}\pi(19)^2(21) = 2,527\pi \approx 7,938.8$  cubic inches.

**12.** 332 square feet   $S = 2B + Ph = 2(2)(8) + 20(15) = 32 + 300 = 332$ square feet. (Alternate method: $S = 2lw + 2wh + 2lh = 2(2)(8) + 2(8)(15) + 2(2)(15) = 32 + 240 + 60 = 332$ square feet.)

**13.** 1,017.9 cubic centimeters   $S = \pi r^2 + \pi rl = 216\pi$ , and $l = r + 6$, so $\pi r^2 + \pi r(r + 6) = 216\pi$. Simplify to $\pi r^2 + \pi r^2 + 6\pi r = 2\pi r^2 + 6\pi r = 216\pi$ or $r^2 + 3r - 108 = 0$ . Solve by factoring. $(r + 12)(r - 9) = 0$ gives a solution of $r = -12$, which is rejected, and $r = 9$. If $r = 9$ then $l = 15$ and $h = 12$. $V = \frac{1}{3}\pi r^2 h = \frac{1}{3}\pi(9)^2(12) = 324\pi \approx 1,017.9$  cubic centimeters.

**14.** 14.7 inches   The area of the equilateral triangle base is $B = \frac{1}{2}(4)(2\sqrt{3}) = 4\sqrt{3}$ inches. $V = \frac{1}{3}Bh$ becomes $34 = \frac{1}{3}(4\sqrt{3})h$. Simplify and solve to find $h = \dfrac{102}{4\sqrt{3}} = \dfrac{102\sqrt{3}}{12} = \dfrac{17\sqrt{3}}{2} \approx 14.7$ inches.

**15.** 7 meters   $S = 2\pi r^2 + 2\pi rh$ becomes $168\pi = 2\pi r^2 + 2\pi r(5)$, which simplifies to $r^2 + 5r - 84 = 0$. Factor to get $(r + 12)(r - 7) = 0$, and solve, rejecting the negative solution and concluding that $r = 7$ meters.

**16.** 400 cubic centimeters   $S = B + \frac{1}{2}lP$ becomes $360 = 100 + \frac{1}{2}l(40)$ or $20l = 260$ and the slant height $l = 13$. Then $V = \frac{1}{3}Bh = \frac{1}{3}(100)(12) = 400$ cubic centimeters.

**17.** The cone has the larger surface area.  For the cylinder, $d = 20$ inches, so $r = 10$ inches and $h = 20$ inches. The surface area of the cylinder $S = 2\pi r^2 + 2\pi rh = 2\pi(100) + 2\pi(200) = 600\pi$ square inches. For the cone, the radius $r = 15$

inches and height $h = 36$ inches, so the slant height is $l = \sqrt{15^2 + 36^2} = \sqrt{1{,}521} = 39$ inches. The surface area $S = \pi r^2 + \pi r l = \pi(225) + \pi(15)(39) = 225\pi + 585\pi = 810\pi$. The cone has the larger surface area.

**18.** 5,652,833 cubic meters    $V = \frac{1}{3}Bh = \frac{1}{3}(65{,}225)(260) = 5{,}652{,}833.\overline{3}$. The volume of the TransAmerica pyramid is approximately 5,652,833 cubic meters.

**19.** 39,219,187.2 cubic feet    $V = Bh = (79{,}515\pi)(157) = 12{,}483{,}855\pi \approx 39{,}219{,}187.2$ cubic feet.

**20.** 11.3 grams per cubic centimeter    The diameter of the cylinder is 6 cm, the radius is 3 cm and the height is 2 cm. The volume of the cylinder is $V = \pi r^2 h = \pi(3)^2(2) = 18\pi \approx 56.5$ cubic centimeters. If cast from iron, the cylinder will have a mass of $18\pi(7.9) \approx 446.7$ grams. Casting it in lead would increase its mass by 192 grams, making the mass of the lead cylinder approximately 638.7 grams. The density of lead is approximately 638.7 grams per 56.5 cubic centimeters, or approximately 11.3 grams per cubic centimeter.

# Posttest

**1.** C    Translation will not change the orientation.

**2.** B    $m\angle RPV = \frac{1}{2}(m\overset{\frown}{RV} - m\overset{\frown}{QS}) = \frac{1}{2}(80° - 20°) = 30°$

**3.** B    Perimeter of $\triangle ADE = AD + DE + AE = 4 + 8 + 7 = 19$

**4.** A    $PQRT$ is a trapezoid but because $\overline{QR} \parallel \overline{PT}$ and $\overline{RS} \parallel \overline{PQ}$, $PQRS$ is a parallelogram, so $m\angle PSR = m\angle Q = 108°$. Because linear pairs are supplementary, $m\angle RST = 72°$. If $\triangle SRT$ is isosceles with vertex $\angle T$, then $TS = TR$ and $m\angle RST = m\angle SRT = 72°$.

**5.** C    Because their bases are squares, the cube and the square pyramid could have square cross-sections. A rectangular prism could possibly have a square cross section if one of its bases is square. The triangular prism would have only triangular cross sections.

**6.** C    $\cos(\angle A) = \sin(\angle C) = \frac{8}{17}$

**7.** D    The exterior angle is 115° and is equal to $m\angle ABC + m\angle BAC$, so A cannot be true. Choice B might be true but there is no information to support it. Choices C and D are similar but only D has the right pair of angles.

**8.** C    To find the length of the legs, use a proportion with the leg as the geometric mean between the hypotenuse $(4 + 12 = 16)$ and the segment of the hypotenuse nearest the side.
$\dfrac{4}{AB} = \dfrac{AB}{16}$ solves to give $AB = 8$, and

$\dfrac{12}{BC} = \dfrac{BC}{16}$ solves to give $BC = \sqrt{192} \approx 13.9$.

Perimeter $\approx 16 + 8 + 13.9 = 37.9$.

**9.** B    Find the length of each side using the distance formula.

$RS = \sqrt{(6-4)^2 + (3-1)^2} \approx 2.8$,

$ST = \sqrt{(4-0)^2 + (1-7)^2} \approx 7.2$,

$RT = \sqrt{(6-0)^2 + (3-7)^2} \approx 7.2$
Perimeter $= 2.8 + 7.2 + 7.2 = 17.2$.

**10.** B    Choice A says the diagonals of the parallelogram are congruent, which is sufficient. Choice C says the parallelogram has a right angle and choice D says that consecutive angles, which are supplementary in any parallelogram, are also congruent, so both right angles. Choice C says the diagonals are perpendicular, which would prove the quadrilateral is a rhombus, but would not assure that it is a rectangle.

**11.** C    The sine of an angle is equal to the cosine of its complement, so $\sin(2x + 7) = \cos(4x - 1)$ can become $\cos(90 - (2x+7)) = \cos(4x - 1)$ and that means that $(90 - (2x+7)) = 4x - 1$. Simplify to get $83 - 2x = 4x - 1$ and solve to get $6x = 84$ and $x = 14$.

**12. B**  $m\angle AEC = \frac{1}{2}\left(m\widehat{AC} + m\widehat{DB}\right) =$
$\frac{1}{2}(51° + x) = 42°$ Solve to find $x = 33°$.

**13. D**  $\overline{FE} \parallel \overline{AB}$ divides sides proportionally, so if $EC = 2\sqrt{3}$, then $BE = \sqrt{3}$. To find DE, use the Pythagorean theorem. $a^2 + b^2 = c^2$ becomes $2^2 + \sqrt{3}^2 = 4 + 3 = 7 = c^2$. $DE = \sqrt{7}$.

**14. B**  The lighthouse, distance out to sea, and line of sight form a right triangle. $\tan(x) = \dfrac{155}{5,280}$

so $x = \tan^{-1}\left(\dfrac{155}{5,280}\right) \approx 1.68°$, which rounds to 2°.

**15. C**  To divide the segment in ratio 2:3, find the points that divide it into 5 equal lengths and then group into two-fifths and three-fifths. Horizontal change is $\dfrac{4-1}{5} = 0.6$ and vertical change is

$\dfrac{-1-9}{5} = -2$. Starting from (1, 9) take 2 steps to $(1 + 2(0.6), 9 - 2(2)) = (2.2, 5)$.

**16. C**  The reflecting line will be the perpendicular bisector of the line segments connecting a vertex to its image. Those would be (2, 7), (4.5, 4.5), and (5, 4). The slope of the line that passes through those points is $m = \dfrac{y_2 - y_1}{x_2 - x_1} = \dfrac{4-7}{5-2} = -1$, so the equation of the line is $y - 7 = -1(x - 2)$ or $y = 9 - x$.

**17. D**  No conclusion can be drawn about congruent triangles because there is no information about congruent sides; however the parallel lines guarantee that the triangles are similar by AA.

**18. A**  $A = \frac{1}{2}ab\sin\angle C = \frac{1}{2}(21)(16)\sin(80°) = 168(0.9848) \approx 165.4$ square centimeters.

**19. D**  $A = \frac{1}{2}bh$ becomes $165.4 = \frac{1}{2}(16)h$. Solve to find that $SV \approx 20.7$ cm.

**20. B**  The central angles of a regular octagon measure 45°, so rotations of multiples of 45° will carry the octagon onto itself.

**21. B**  Rotating a right triangle about its hypotenuse will not produce a cone, so eliminate choice C. For the remaining choices, determine the radius and height of the resulting cone. In choice A, the radius is 24 and the height is 18. $V = \frac{1}{3}\pi r^2 h = \frac{1}{3}(24)^2(18) = 3,456\pi$.

In choice B, $r = 18$ and $h = 24$, and $V = \frac{1}{3}\pi r^2 h = \frac{1}{3}(18)^2(24) = 2,592\pi$. Choice D requires first finding the length of the altitude (14.4) and the lengths of the two segments of the hypotenuse created by the altitude (10.8 and 19.2). Luckily, choice B is correct.

**22. D**  The endpoints of the segment are (0, −3) and (8, 3). The slope of this segment is
$m = \dfrac{y_2 - y_1}{x_2 - x_1} = \dfrac{3+3}{8-0} = \dfrac{6}{8} = \dfrac{3}{4}$ so the slope of the perpendicular bisector is $-\dfrac{4}{3}$. The midpoint of the given segment is $\left(\dfrac{0+8}{2}, \dfrac{-3+3}{2}\right) = (4, 0)$.

The equation of the perpendicular bisector is $y = -\frac{4}{3}(x - 4)$. Choice A is the midpoint so that lies on the bisector. $y = -\frac{4}{3}(1 - 4) = 4$ and $y = -\frac{4}{3}(7 - 4) = -4$, so B and C are points on the perpendicular bisector. $y = -\frac{4}{3}(3 - 4) = -\frac{4}{3}(-1) = \frac{4}{3} \neq 1$. Choice D is the point that does not lie on the perpendicular bisector.

**23. D**  To find the population (number of people), multiply $\dfrac{\text{people}}{\text{square mile}} \times$ square miles. Results are: New Hampshire 1,325,044; Indiana 6,591,984; Louisiana 4,666,032; Georgia 10,179,801.

**24. A**  To prove the triangles congruent, a third piece of information is necessary, specifically, a pair of congruent angles, to complete AAS. That eliminates choices B and C. To choose between A and D, look at the congruence statement $\triangle ABC \cong \triangle XZY$. $\angle A$ corresponds to $\angle X$. Choice A is correct.

**25. C**  The diagonals of a rhombus divide the rhombus into 4 right triangles, each with legs of 12 and 4.5. Use the Pythagorean theorem to find the hypotenuse of the triangle, which is a side of the rhombus. $a^2 + b^2 = (12)^2 + (4.5)^2 = 144 + 20.25 = 164.25 = c^2$ so $\sqrt{164.25} \approx 12.8$. The perimeter of the rhombus is $4(12.8) = 51.2$ centimeters.

**26. D**  The equation of a circle centered at $(h, k)$ with radius $r$ is $(x - h)^2 + (y - k)^2 = r^2$. Eliminate B and C, which equal $r$, not $r^2$. The center is $(3, -4)$ so $(x - h)^2 + (y - k)^2 = r^2$ becomes $(x - 3)^2 + (y - (-4))^2 = 49$ or $(x - 3)^2 + (y + 4)^2 = 49$. Choice D is correct.

**27.** B  Find the third angle of each triangle, assure that there is a correspondence, and make sure that the similarity statement spells out the correct correspondence.

**28.** B  If the diameter is 30 cm, then the radius is 15 cm. The measure of $\angle AOB$ is equal to the measure of $\overset{\frown}{AB}$ The area of the sector is calculated by $\dfrac{m\angle AOB}{360°} \cdot \pi (15)^2$ and it is given that $\dfrac{m\angle AOB}{360°} \cdot \pi (15)^2 = 20\pi$. Simplify to $\dfrac{225 \cdot m\angle AOB}{360} = 20$ and solve to find

$$m\angle AOB = \frac{20 \cdot 360}{225} = \frac{7{,}200}{225} = 32°.$$

The length of $\overset{\frown}{AB}$ is $\dfrac{32°}{360°} \cdot 2\pi \cdot 15 = \dfrac{4}{45} \cdot 30\pi = \dfrac{8\pi}{3} \approx 8.4$ cm.

**29.** B  The intersection points C and D define the perpendicular bisector because they are equidistant from A and B. All that remains is to connect them by drawing $\overleftrightarrow{CD}$.

**30.** C  Choices B and D are not true. Choice A is true, but not enough to ensure that the line is the perpendicular bisector.

**31.**

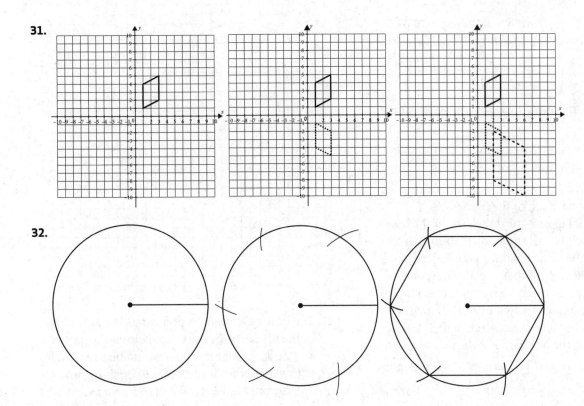

**32.**

**33.** If *ABCD* is an isosceles trapezoid (given), then $\overline{AB} \cong \overline{CD}$ by the definition of isosceles trapezoid. If *F* is the midpoint of $\overline{BC}$, then $\overline{BF} \cong \overline{FC}$ by the definition of midpoint. $\angle B \cong \angle C$ because base angles of a trapezoid are congruent. $\triangle ABF \cong \triangle DCF$ by SAS, and $\overline{AF} \cong \overline{FD}$ by CPCTC. Given that $\angle EAD \cong \angle EDA$, by the Base Angle Converse theorem, $\overline{AE} \cong \overline{ED}$. $\overline{FE} \cong \overline{FE}$ (reflexive) and $\triangle AFE \cong \triangle DFE$ by SSS.

**34.** First, find the midpoint of the diagonals. In a parallelogram, the diagonals bisect each other, so expect them to have the same midpoint, which we'll label *X*.

$$X = \left( \frac{d + (a + c)}{2}, \frac{e + b}{2} \right)$$
$$= \left( \frac{a + (d + c)}{2}, \frac{b + e}{2} \right)$$
$$= \left( \frac{a + c + d}{2}, \frac{b + e}{2} \right).$$

Next, find the midpoints of opposite sides of the parallelogram, which we'll label $M$ and $N$.

$$M = \left(\frac{a+d}{2}, \frac{b+e}{2}\right) \text{ and}$$

$$N = \left(\frac{(a+c)+(d+c)}{2}, \frac{b+e}{2}\right)$$

$$= \left(\frac{a+2c+d}{2}, \frac{b+e}{2}\right)$$

Now find the equation of the line that connects $M$ and $N$, starting with the slope.

$$m = \frac{y_2 - y_1}{x_2 - x_1} = \frac{\dfrac{b+e}{2} - \dfrac{b+e}{2}}{\dfrac{a+2c+d}{2} - \dfrac{a+d}{2}} = \frac{0}{c} = 0$$

The slope of the line is zero, so it is a horizontal line and has the form $y = \dfrac{b+e}{2}$. Point $X$

has a $y$-coordinate of $\dfrac{b+e}{2}$, so it is a point on

$\overline{MN}$. $\overline{MN}$ passes through the midpoint of both diagonals, so it bisects both diagonals.

35. Smallest to largest: square pyramid, hexagonal pyramid, cone. Each of the figures has a volume equal to one-third the area of the base times the height. All three calculations have the $\frac{1}{3}$ and the same height, so the difference will be in the area of the base. For the cone, $B = \pi r^2$. In the hexagonal pyramid, each of the small triangles created by central angles is equilateral, because the central angle is 60°. If the radius of the hexagon is $r$, the side is $r$, and the apothem is $\frac{1}{2}r\sqrt{3}$. The area of the hexagon is $B = \frac{1}{2}aP = \frac{1}{2}\left(\frac{1}{2}r\sqrt{3}\right)(6r) = \frac{3}{2}r^2\sqrt{3}$. For the square, if the radius is $r$, the side is $r\sqrt{2}$ and the area of the square base is $B = (r\sqrt{2})^2 = 2r^2$. Comparing $\pi r^2$, $\frac{3}{2}r^2\sqrt{3}$, and $2r^2$ means ordering $\pi$, $\frac{3}{2}\sqrt{3}$, and 2. Because $\pi \approx 3.14$ and $\frac{3}{2}\sqrt{3} \approx 2.59$, we can conclude the square pyramid has the smallest volume $\left(\frac{1}{3} \cdot 2r^2h = \frac{2}{3}r^2h\right)$, the hexagonal pyramid $\left(\frac{1}{3} \cdot \frac{3}{2}r^2\sqrt{3}h = \frac{1}{2}\sqrt{3}r^2h\right)$ ranks second, and the cone $\left(\frac{1}{3}\pi r^2h\right)$ has the largest volume.

36. Circle with a radius of 9, centered at $(-6, 3)$. The equation $x^2 + 12x + 36 + y^2 - 6y + 9 = 81$ does not immediately look like the equation of a circle, but if you compare it to the form you expect, it can be transformed.

$$(x - h)^2 + (y - k)^2 = r^2$$

$$(x^2 + 12x + 36) + (y^2 - 6y + 9) = 81$$

A little factoring will convert the given equation to $(x + 6)^2 + (y - 3)^2 = 9^2$. The equation represents a circle with a radius of 9, centered at $(-6, 3)$.

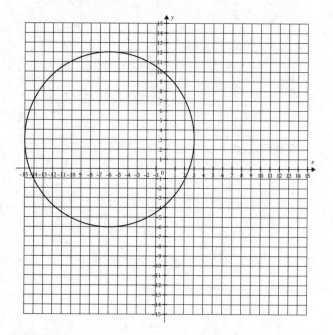

37. The best location for a pool would be at the rear of the property, in the space behind the garage and deck. The largest pool would be a rectangle with vertices at (50, 20), (65, 20), (65, 75), and (50, 75). It would have an area of 825 square feet. The best circular option would be centered at (50, 55) with a radius of $5\sqrt{10} \approx 15.8$ feet. The area would be slightly over 784 square feet. A circle centered at (55, 55) with a radius of 15 would give an area of almost 707 square feet, and a rectangle with vertices (35, 45), (65, 45), (65, 65) and (35, 65) would have an area of 600 square feet.

38. The building covers an area of approximately 115,849 square meters. The side of the Pentagon is 281 meters, so its perimeter is 1,405 meters. To find the apothem, focus on the isosceles triangle created by one central angle.

The central angle will measure $360° \div 5 = 72°$, so the base angles of the triangle will be $\frac{180° - 72°}{2} = \frac{108°}{2} = 54°$. Half of a side

will be 140.5 meters. $\tan(54°) = \dfrac{a}{140.5}$, so

$a = 140.5\tan(54°) \approx 193.38$ meters. Then

$A = \frac{1}{2}aP = \frac{1}{2}(193.38)(1,405) = 135,849.45$ square meters. This is the area enclosed by the outer wall of the building, including the courtyard. Subtract the 20,000 square meters of courtyard, and the building covers an area of approximately 115,849 square meters.

The side of the courtyard is about 108 meters long. To find the side of the courtyard, express all the quantities in the area formula in terms of the side length. The apothem

can be found from $\tan(54°) = \dfrac{a}{s/2}$ or

$a = \dfrac{s}{2}\tan(54°)$. Perimeter is easier: P = 5s. Then

$A = \frac{1}{2}aP = \frac{1}{2}\left(\frac{s}{2}\tan 54°\right)(5s) = \dfrac{5s^2}{4}\tan(54°)$ and

we know that the area of the courtyard is 20,000 square meters. Solve:

$$\dfrac{5s^2}{4}\tan(54°) = 20,000$$
$$s^2\tan(54°) = 16,000$$
$$s^2 = \dfrac{16,000}{\tan(54°)} \approx 11,624.68$$
$$s \approx \sqrt{11,624.68} \approx 107.8$$